CHIMPANZEE POLITICS

CHIMPANZEE

REVISED EDITION

with photographs and drawings by the author

Power and Sex among Apes

POLITICS

Frans de Waal

The Johns Hopkins University Press   Baltimore and London

© 1982, 1989, 1998 Frans de Waal
All rights reserved. Revised edition, 1998
Printed in the United States of America on acid-free paper
9 8 7 6 5 4 3 2 1

Hardcover edition first published in the United States in 1982
by Harper & Row, Publishers, Inc., New York; Johns Hopkins
Paperbacks edition published in 1989.

The Johns Hopkins University Press
2715 North Charles Street
Baltimore, Maryland 21218-4363
The Johns Hopkins Press Ltd., London

Library of Congress Cataloging-in-Publication Data will be found
at the end of this book.

A catalog record for this book is available from the British Library.

ISBN 0-8018-5839-9

For Jan van Hooff

I put for a generall inclination of all mankind,
a perpetuall and restlesse desire of Power after power,
that ceaseth onely in Death.

THOMAS HOBBES, 1651

CONTENTS

Gallery of color photographs follows page 104.

PREFACE TO THE REVISED EDITION

WHEN I WROTE *CHIMPANZEE POLITICS*, IN 1979 AND 1980, I WAS A beginning scientist, in my early thirties, without much to lose. At least, that's the way I looked at it at the time. I didn't mind following my intuitions and convictions, however controversial these might be. Keep in mind that this was a time at which the words *animal* and *cognition* could barely be mentioned in the same sentence without raising eyebrows. Most of my colleagues shied away from the suggestion of intentions and emotions in animals for fear of being accused of anthropomorphism. Not that they necessarily denied animals an inner life, but they followed the behaviorist dogma that, since what animals think and feel is unknowable, there is no point in talking about it. I still remember standing for hours on the metal grid over the smelly night quarters of the chimpanzees, holding the only phone in the building to my ear, talking with my professor, Jan van Hooff, who, though always supportive, was also quite a bit more cautious than I, trying to convince him of yet another wild speculation. It is during these discussions that Jan and I, at first jokingly, began referring to developments in the colony as "politics."

The other major influence on this book was the general public. For years, I addressed organized groups of zoo visitors, including lawyers, housewives, university students, psychotherapists, police academies, bird-watchers, and so on. There is no better sounding-board for a would-be popularizer. The visitors would yawn at some of the hottest academic issues, but react with recognition and fascination to basic chimpanzee psychology that I had begun to take for granted.

I learned that the only way to tell my story was to bring the chimpanzee personalities to life and to pay attention to actual events rather than the abstractions that scientists are so fond of. I benefited greatly from a previous experience. Before I came to Arnhem, I had done a dissertation project at the University of Utrecht. In one of my monkey groups, the males had changed ranks, resulting in my very first scientific paper, published in 1975, entitled: *The wounded leader: A spontaneous temporary change in the structure of agonistic relations among captive Java-monkeys*. In putting this report together, I had noticed how utterly useless the cus-

My eyes were not the only ones riveted on the drama in the colony: the apes, themselves, kept a close watch as well. A few of them look on while Nikkie (background, left) is waking up Yeroen with an intimidation display.

tomary formalized records of ethologists are when it comes to social drama and intrigue. Our standard data collection aims at categorizations that serve the counting of events. Computer programs sort through the data, creating neat summaries of aggressive incidents, grooming bouts, or whatever behavior we are interested in.

Items that cannot be quantified and graphed run the risk of being tossed aside as mere "anecdotes." Anecdotes are unique events from which it is hard to generalize. But does this justify the contempt in which some scientists hold them? Let's consider a human example:

Bob Woodward and Carl Bernstein describe in *The Final Days* Richard Nixon's reaction to his loss of power: "Between sobs, Nixon was plaintive. . . . How had a simple burglary done all this? . . . (He) got down on his knees . . . leaned over and struck his fist on the carpet, crying aloud, 'What have I done? What has happened?'"

Nixon was the first and only U.S. president to resign, so this really can't be much else than an anecdote. But does this diminish the observation's significance? I must admit to a great weakness for rare and peculiar events. As we shall see, one of my chimpanzees had tantrums similar to Nixon's (minus the words) under similar conditions. I learned from my earlier study that in order to analyze and understand such events one needs a diary that conveys how things unfolded, how each individual got involved, and what was special about a situation compared to previous ones. Instead of merely counting up and averaging chimpanzee behavior, I was intent on injecting historiography into my project.

Thus, upon arrival in Arnhem I opened a diary. Since little happened initially, I filled it with notes about personalities and behavior patterns that struck me as unusual. As a result, however, I was getting into a chronicling mode, sensitive to shifting social relationships, ready for the political drama that was to come. When at last it did explode, I filled page after page with impressions, predictions, corrections of earlier impressions, but most of all the bare facts. Fascinated and emotionally engaged, I spent day-in-day-out and thousands of hours on a wooden stool overlooking the island, intent on producing the most detailed record ever of a power struggle, human or nonhuman. It is only in sifting through my copious notes, years later, that the connections between various events fell into place, and *Chimpanzee Politics* began to take shape.

The book caused little controversy when it first appeared, in 1982, with the publishing house of Jonathan Cape, in London. In both popular and academic reviews it was welcomed rather than attacked. In hindsight, this is understandable as its underlying premise perfectly fit the *Zeitgeist* of the 1980s in which attitudes toward animals were rapidly changing. Having worked largely in isolation from the emergence of cognitive psychology in America, I had not realized that I had not been alone in the exploration of this new intellectual territory. This circumstance illustrates how scientific developments in different corners of the world are often connected by thin threads of shared ideas. They are never totally independent. Thus, Donald Griffin's *The Question of Animal Awareness* did not surprise me when I first read it, just as *Chimpanzee Politics* evidently did not surprise most primatologists.

If we follow Harold Laswell's famous definition of politics as a social process determining "who gets what, when, and how," there can be little doubt that chimpanzees engage in it. Since in both humans and their closest relatives the process involves bluff, coalitions, and isolation tactics, a common terminology is warranted. The title of my book drove this point home. Whereas several political scientists had no objections, one among them felt a need to delineate humans as quite different.[1] As so often in the history of ape-human comparisons, ad hoc changes were made in the definition of a phenomenon so as to exclude the primate data. At a symposium on the application of political theory to animals, Glendon Schubert proposed that the term "politics" be reserved for processes within groups of at least one hundred individuals without kinship ties. This obviously excluded most social animals, as well as many situations in which humans play power games.

Chimpanzee Politics was written with a general audience in mind, but it also found its way into the classroom and to business consultants, and even became recommended reading for freshman congressmen. Because of the undiminished interest over a fifteen-year period, the Johns Hopkins University Press and I decided that it was time for a revised edition. There can, of course, be no rewriting of history: my original account of the power struggles has been left untouched. Rather, I have modified the text in view of new knowledge, added notes to highlight the latest research, reprinted all black-and-white photographs so as to reproduce them at optimal quality, added new photographs, both black & white and color (most of which have never been published before), and appended an Epilogue detailing subsequent developments in the Arnhem colony. The result, I hope, is a more attractive and updated version of the original book.

In explaining the insights gained from my study, I am tempted to draw a parallel with island biogeography. Ecological complexity increases with the number of species of plants and animals. Since islands usually have a smaller variety of species than the mainland, study of their flora and fauna has greatly helped clarify rules of extinction and survival as well as other basic ecological principles. The relative simplicity of islands has allowed naturalists, from Charles Darwin to Edward

Wilson, to develop ideas applicable to more complex systems. Similarly, the chimpanzee island at the Arnhem Zoo housed a limited number of chimpanzees, under simplified conditions compared with an equatorial rain forest. Imagine that the number of male players in the colony had been three times as high—as it often is in wild communities—or that the chimpanzees had been free to move on and off the island. I probably would not have been able to make much sense of the drama that was acted out in front of me. Like an island biogeographer, I saw more because there was less. Yet, the general principles that I uncovered apply not only to apes on an island but to jockeying for power everywhere.

CHIMPANZEE POLITICS

INTRODUCTION

Visitors to a zoo always appear amused by the sight of chimpanzees. No other animal attracts so much laughter. Why should this be? Are they really such clowns, or does their appearance make them ridiculous? It is almost certainly their looks that amuse us, because they need do little more than walk around or sit down to make us laugh. The hilarity is perhaps a camouflage for quite different feelings—a nervous reaction caused by the marked resemblance between humans and chimpanzees. It is said that apes hold up a mirror to us, but we seem to find it hard to remain serious when confronted with the image we see reflected.

It is not only visitors to the zoo who are fascinated but uneasy in the presence of chimpanzees; the same is true of scientists. The more they learn about these great apes, the deeper our identity crisis seems to become. The resemblance between humans and chimpanzees is not only external. If we look straight and deep into a chimpanzee's eyes, an intelligent, self-assured personality looks back at us. If they are animals, what must we be?

A whole series of facts are now known which reduce the gap between humans and animals. Gordon Gallup has proven that great apes recognize themselves in a mirror. This form of self-awareness seems to be lacking among monkeys and other animals, who regard their reflection as if it were someone else. Wolfgang Köhler carried out ingenious intelligence tests on chimpanzees and concluded that they are capable of solving new problems on the basis of a sudden realization of cause and effect (the "aha! experience"). Jane Goodall saw wild chimpanzees using self-made tools. They were also seen to hunt, eat meat, extend their territory by means of "warfare" and even to be capable of cannibalism. Finally, the husband-and-wife team of R. Allen Gardner and Beatrice Gardner succeeded in teaching chimpanzees a large number of symbols, in the form of hand gestures, which they used to communicate in a manner surprisingly similar to the way we use language. These apes revealed a wealth of information about what they were thinking and feeling: the ape mind was made accessible to our species.

But however impressive all these discoveries may be, one important link is still missing: the social organization. There is evidence that chimpanzees lead a highly subtle and complex social life, but the picture is

Krom (left) *and Gorilla grooming each other.*

still disjointed. Up to now research into this particular field has been carried out almost exclusively with wild chimpanzees. These observations are extremely important, but it is impossible to follow social processes in every detail in the jungle. Fieldworkers are lucky if they so much as see the animals regularly. Out of the thousands of social contacts that take place in the undergrowth and in trees, they will witness only a few. They will not fail to note the results of social changes, but they will often be ignorant of the causes.

There is at present only one place in the world where a comprehensive study of the group life of these fascinating animals is possible: the large, open-air chimpanzee colony at Burgers Zoo in Arnhem. Such a study has been going on for some years now. This book presents the results and demonstrates something we had already suspected on the grounds of the close connection between apes and humans: that the social organization of chimpanzees is almost too human to be true. The clowns of the animal world would obviously feel very much at home in a political arena. Entire passages of Machiavelli seem to be directly applicable to chimpanzee behavior. The struggle for power and the resultant opportunism is so marked among these creatures that a radio reporter once thought to try and surprise me with the question: "Who do you consider to be the biggest chimpanzee in our present government?"[2]

Every day the newspapers administer large doses of political commentary. We are used to political developments being outlined for us in neat generalizations, such as "Split in government's camp plays into opposition's hands" or "Minister puts himself in an impossible position." Political journalists frequently do not enumerate the many factors and incidents which have led to this situation. No one expects them to go into exhaustive detail about all the political statements that have been made and all the confidential information they have gleaned. By and large their readers are satisfied with the general outline.

The events I witnessed in Arnhem could also be summarized in this way. It would certainly be the easiest way of talking about them, but the picture I would sketch would lack conviction. My interpretations are inevitably regarded with more suspicion than the interpretations of a political journalist. The term *politics* in itself gives rise to doubts where animals are concerned.

That is why I feel bound to approach the subject step by step, beginning with an outline, in this introduction, of what chimpanzee communication is all about. The first chapter then provides a character sketch of the group members. Subsequent chapters report on the various bids for power that we have witnessed during the six years in which the project was in operation and how rank reversals affect sexual privileges. Finally, I discuss some general mechanisms underlying social interaction—such as reciprocity, strategic intelligence and triadic awareness—and point out how similar they are to human mechanisms.

First Impressions

Once inside the gate of Arnhem Zoo, visitors find themselves wandering down the oldest and widest avenue in the park. On their way they pass the parrots, pelicans, and flamingoes on the left and the parakeets, owls, and pheasants on the right. About halfway down this avenue, above the cacophony of bird noises, a much more raucous shouting can be heard. The shouts come from the chimpanzees in their huge open-air enclosure at the end of the avenue.

Having reached this point the visitors may be disappointed to find the apes are still some 20 meters away, so as to deter the public from trying to feed them. If they want to watch the animals more closely, they will need to go up into the lookout post. From behind unbreakable glass (chimpanzees pelt onlookers with stones!) they will have a splendid view of the whole outdoor enclosure, which covers almost two acres. It is surrounded by a wide moat filled with water. This area used to be part of a large wood, and there are still about fifty tall oak and beech trees on the island, most of which are protected by electric fencing against the destructive habits of the island's occupants. Some oak trees have been left unprotected, and they can be seen standing in the center of the enclosure, completely stripped. These dead oaks play an important role in the life of the group. Major aggressive encounters always end in the tops of these trees, which offer numerous possibilities for evading adversaries.

Some members of the public clearly still have to get used to the new, seminatural layout. Opportunities for feeding, touching, and provoking the apes have been reduced to virtually zero. The only thing the visitors can do is stand and watch. But the great advantage is that there is a great deal more to see here than in classic ape houses, where two to four chimpanzees share a cramped and uninspiring cage. In such degrading conditions apes often do little more than lie around masturbating boredly,

Top, *overview of the Arnhem Zoo chimpanzee exhibit. On the right is the
building containing the night quarters and indoor halls for the winter. On the left
is the wall that the chimpanzee once conquered. Drawing by Bonnie Willems.*
Bottom, *part of the open-air enclosure with the dead oak trees in the middle.*

pace up and down, or bump their backs or even their heads rhythmically against the wall of their cage.[3]

Visitors will not find any of these behavior patterns in the Arnhem colony. The most frequent social activity is a perfectly natural one: grooming. Several apes are usually to be found assembled in grooming clusters, where they attend to one another's hair. Soft spluttering and smacking sounds accompany this meticulous work, and every now and again the grooming partner is gently pushed or pulled into a new position. The willingness with which the instructions are followed indicates just how much chimpanzees enjoy being groomed.

When adult females form a grooming cluster their children are usually to be seen wandering in the vicinity, while the very tiny ones sit firmly and safely clamped to mother's tummy and watch everything going on around them. The slightly older children seem to be endowed with inexhaustible energy. When they play tag they rush straight through the middle of the grooming clusters, disturbing the full-grown apes by jumping on top of them or throwing handfuls of sand.

A group of relaxing apes. Jimmie (left) is grooming Tepel. Jimmie's youngest child is sitting in between them. In the center are the five-year-old sons of the two females: Wouter is tickling Jonas under his armpit. Krom is sitting on the right.

Only in a harmonious group are adult males solicitous toward the children and tolerant of their behavior: above, Moniek quite happily allows herself to be lifted up in the air during one of her frequent games with Nikkie; right, Luit allows his back to be used as a trampoline.

The Arnhem colony is unique not only in the vastness of the open-air enclosure and the large number of young growing up with their mothers but above all in its size (approximately twenty-five individuals) and in the fact that there are several adult males living in the group.

The males are not very much bigger than the females, but they have a thicker coat. When they are excited or aggressive, their hair stands on end so that they appear larger than life and frighteningly impressive. At such moments chimpanzee males can be amazingly quick on their feet. These aggressive turns are often announced a good ten minutes before by inconspicuous body movements and changes of posture. When I am showing visitors round and I notice the signs of an imminent intimidation display, I myself get a chance to impress, in the typically human form of showing off my knowledge. I have ample time to predict to any unsuspecting guests what scenes they are about to witness.

The predictability of chimpanzee behavior does not mean, however, that they always repeat the same social patterns. That would be boring. The most fascinating aspect of studying chimpanzees is recording the changes which take place over the years. Making short-term predictions is not only fun as a means of surprising other people, but it is also an extremely useful way of constantly checking my knowledge of the ever-changing system of relationships within the group.

The dynamism of the group's life is most clearly illustrated by the leadership changes which have occurred in the Arnhem colony. These processes took many months and, contrary to what people so often think, they were not decided by a few fights. My own research has been particularly concerned with the endless series of unobtrusive social maneuvers leading up to the dethronement of the leader. The stability of the group is slowly undermined. Each individual has his or her role to play in this web of intrigues. The future new leader shows the way, but he can never act entirely alone; he cannot impose his leadership upon the group single-handed. His position is granted him, in part, by the other chimpanzees. The leader, or alpha male, is just as much ensnared in the web as the rest.

Prevention of Explosive Tension

For many years now zoos have kept monkey species such as baboons and macaques in fairly natural groups on the familiar monkey rocks.

But for the great apes there has been no such congenial group life. Zoo owners feared that a large colony of these frightening and unpredictable creatures would lead to bloody clashes and even death. What is more, great apes are extremely susceptible to disease, and it was hoped that by isolating the animals in sterile cages the danger of infection would be eliminated. In 1966, however, the brothers Anton and Jan van Hooff decided to attempt an ambitious project at Arnhem Zoo. Jan was able to benefit from the experience he had gained in America, studying the social behavior of chimpanzees in a vast colony at Holloman Air Force Base in New Mexico. There the chimpanzees were living together in a 25-acre open-air enclosure.

The idea behind the American colony was excellent, and yet it had not been a success. There was an extremely tense and aggressive atmosphere in the group. Jan deduced that the major error was the lack of facilities for separating the apes at feeding time. Violent fights broke out at every meal because some of the apes tried to monopolize the food. The tension began to build up long before feeding time. This meant that one of the basic prerequisites for the development of a harmonious group life was missing.

In their natural environment chimpanzees forage for food on their own or in small groups. The fruits and leaves they are searching for are so evenly scattered that competition for food is unusual. But as soon as humans start providing food, even in the jungle, the peace is quickly disturbed. This happened in Gombe Stream in Tanzania, where Jane Goodall carried out her famous studies. Richard Wrangham concluded that by systematically feeding bananas to the chimpanzees in Gombe the aggression rate increased sharply.

In Arnhem the problem of competition for food has been effectively solved by two measures. First, the public are kept away from the animals so that they cannot feed them. Second, the apes are split up every evening into small groups and fed in the ten cages where they sleep. They rarely eat while they are with the whole group; they each receive their own fair share in their cages every evening and every morning. Their diet consists of apples, oranges, bananas, carrots, onions, bread, milk, and sometimes an egg. Their staple food is compressed food pellets (monkey chow), which contain carbohydrates, proteins, and vitamins. During the summer the chimpanzees eat large quantities of grass plus acorns, beech nuts, leaves, insects, and some edible mushrooms.

A playful contest between Tarzan (left) *and Jonas.*

In order to get enough to eat, wild chimpanzees have to spend more than half their time foraging. Since they do not need to do this in a zoo, they will inevitably be slightly bored. The result is that their social life becomes intensified. They have more than enough time to "socialize." In addition, their quarters are limited, so they can never completely isolate themselves from the group. The results are especially marked during the winter months.

The Dutch winters (from late November to mid April), which are severe for chimpanzees, are spent in a heated building containing their sleeping quarters and two large halls with climbing frames and hollow metal drums. (The adult males give noisy, rhythmical concerts on these drums.) The largest hall is 21 meters long and 18 meters wide. Although this may seem reasonable, it is only one-twentieth of the size of the open-air enclosure. This gives rise to irritation and friction; aggressive incidents are nearly twice as common in winter as they are in summer.

The day the chimpanzees leave their winter quarters is the most fes-

Early in the morning Zwart walks over, on two legs because the grass is still wet, to join a group including Amber (right). *She is greeted with a playful slap from Moniek.*

tive day of the year. In the morning the keeper opens the trapdoor leading to the open-air enclosure. The apes cannot see what is going on from their sleeping quarters, but they can distinguish all the trapdoors in the building by ear. Within a second the entire colony reacts with a deafening scream. They are let out into the open air in small groups. The screaming and hooting continues. All over the enclosure apes can be seen embracing and kissing each other. Sometimes they stand in groups of three or more jumping and thumping each other excitedly on the back.

The apes' delight in regaining their freedom is obvious. Their black coats, which have grown thin during winter, will become thick and shiny again within a few months. Pale faces will regain their color in the sun. And, most important of all, the tension, which has been bottled up all winter, will dissolve in the open air.

The Great Escape

The existence of our primate exhibit is due to the enterprise and daring of its director, Anton van Hooff, and his philosophy that it is better for a zoo to house a few animal species well than many species badly. In August 1971 the complex was officially opened by Desmond Morris. Surrounded by other impeccably dressed Naked Apes he pronounced the opening words, after which our hairy relatives were let out into the open-air enclosure. Our guest speaker predicted that one of two disasters would befall us in due course: either the apes would construct a raft and get across the moat, or they would invent a ladder and use it to scale one of the walls of the enclosure. The first disaster he had thought up himself, the second was a reference to a discovery made by Rock, a chimpanzee.

Rock was the oldest of a small group of juvenile chimpanzees in Louisiana who were being studied by Emil Menzel. Completely on his own Rock had hit on the brilliant idea of using a long pole as a kind of ladder to climb over a wall. The other chimpanzees in the group quickly grasped the use of this instrument. They even helped each other in their climbing efforts.

The most memorable escape in the history of the Arnhem colony took place in a similar way. Despite the warning we had been given during the opening, several large branches were left lying around on the apes' island. A small section of the enclosure is bounded by a wall 4 meters high. The story of what happened has become a classic at the zoo. According to the most popular version, the chimpanzees placed branches against a wall at different points and simultaneously scaled the wall as if the plan had been agreed beforehand. It resembled the storming of a medieval castle, with the chimpanzees all helping one another over the ramparts. Then more than ten chimpanzees made a bee-line for the large restaurant, which was crowded out at the time. There they ate their fill of oranges and bananas, after which they ambled back to their sleeping quarters, their hands and feet crammed with stolen fruit, and spent the rest of the day eating to their hearts' content.

After years of having heard this splendid story I was somewhat disappointed when I checked up here and there for details, with this book in mind. I asked everyone what they had seen with their own eyes. The outcome was predictable. The tale contained a nucleus of truth, but it had been fairly freely embroidered on in the intervening years. The restaurant staff, for example, told me that they had never stocked fruit and that on the day of the escape only one chimpanzee had in fact come in. This was Mama, the oldest and undoubtedly also the most dangerous female in the group. She had apparently climbed over the counter and investigated the cash desk before settling down in the midst of a group of visitors and quietly emptying a bottle of Chocomel.

I did not talk to anyone who had witnessed the actual breakout. That it was effected with the aid of a branch is certain (a heavy branch 5 meters long was found propped up against the wall), but whether several such branches were used simultaneously remains unclear. That the breakout should have been a team effort does not surprise me in the least; the weight of the branch alone makes this likely.

Although the keeper assiduously inspects the chimpanzees' enclosure for broken branches every morning—a practice adopted since the memorable Great Escape—this has not curbed the apes' ingenuity. Finding no loose branches lying around, they break off long sections from the dead oak trees. This requires enormous strength, so the task always falls to the grown males. To our relief the branches are no longer used for escape purposes, but to clamber over the electric fencing in order to get into the live trees.

With such intelligent animals as chimpanzees it is never possible to eliminate every opportunity for escape. They even know how to use keys and sometimes try to fish them out of the keeper's pocket. Escapes are only funny to recount in retrospect. At the time there is nothing to laugh about; all anyone can think of is the danger.

None of us dares to go in among the chimpanzees. Their keeper and I are on very friendly terms with some of them, but only when they are in their sleeping quarters and there are bars between us. Zoos make it a rule never to trust any adult chimpanzee fully. They are no heavier than humans, but they are *much* stronger. The problem with chimpanzees in a zoo is that they are only too well aware of their superior strength. This plus their temperamental nature makes them deadly.

Wild chimpanzees are not aware that they are stronger than humans and, what is more, they have learned to fear humans and their weapons. This results in the paradoxical situation that wild chimpanzees, once they have become used to humans, can be studied at closer quarters than our chimpanzees at Arnhem. We observe them from across the moat at a distance varying between 6 to 60 meters (for the public this distance, except in the lookout post, is even greater). In Gombe, on the other hand, the fieldworkers sometimes simply go and sit down by the chimpanzees and look on. But even in Gombe, chimpanzees have now known people long enough to have lost their timidity. The most notorious character is Frodo, a muscular young adult, who freely slaps around human visitors to the camp, and sometimes drags them down the slopes. During one attack, he almost broke Jane Goodall's neck when he stamped her with full force on the head. He seems to want to dominate and intimidate. There is little the investigators can do to discourage such behavior without damaging hard-won trust.

Ethology

A young teacher brought his class to look at the chimpanzees. It was in the middle of winter and consequently the colony was indoors. Sev-

Wouter, Tarzan, and, behind them, Zwart watch curiously to see what Nikkie has fished out of the moat.

eral apes were sitting and lying around on the tall drums in a corner of the hall. The drums are of different heights and the teacher immediately saw the educative value of the arrangement. The ape sitting on the highest drum, he told his pupils, was the leader of the pack. Below him sat his adjutant and below him their subordinates. In his desire to make everything simple and explicit he also pointed to the "lowliest" apes, namely those sitting and walking around on the ground.

Among the apes on the ground was Yeroen, one of the dominant males, who to my great delight was in the process of warming up for a bluff display. His hair was already standing slightly on end and he was hooting quietly to himself. When he stood up his hooting became louder and some of the apes hurried off the drums, knowing that Yeroen's displays generally ended there with a long, rhythmic stamping concert. I was curious to see how the young teacher would extricate himself from the situation. After Yeroen had produced his customary din and had made several wild charges through the hall, things quietened down again. The other chimpanzees climbed back on to the drums and resumed their activities. The teacher's comment was the product of a rich imagination. The performance they had just witnessed, he told his students, was an abortive attempt by the ape on the ground to seize power.

This was a ridiculous suggestion. But who can guarantee that the many interpretations in this book are in fact the truth? Although I feel that after all these years I know the group intimately and am seldom wrong about the events that take place in it, I cannot be absolutely sure. To study animal behavior is to interpret, but with a constant gnawing feeling that the interpretation may not be the right one. This is not a pleasant sensation, and it is the reason why scientists often prefer to remain silent rather than answer the all-too-familiar question: "Why is that animal doing what it is doing?" Experts sometimes choose to create the impression of knowing nothing. They act in exactly the opposite way from the young teacher, who held forth with such conviction. Both attitudes lead nowhere, but unfortunately I will not be able to avoid them completely. At some points I may seem exaggeratedly hesitant and at others I may appear to go too far in my interpretation. There is no other way. Behavioral study is conducted on a see-saw between these two extremes.

Ethology is biological behavioral study. It gained ground in the 1930s

Observers concentrate on a particular form of behavior or follow a specific individual. Their work is more strenuous than might appear.

in Germany, the Netherlands, and England under the influence of Konrad Lorenz and Niko Tinbergen. The greatest difference between ethology and psychological behavioral study of animals lies in ethology's strong emphasis on *spontaneous* behavior in the *natural* environment, or at least under the most natural conditions possible. Ethologists do conduct experiments, but never completely detached from their fieldwork. They are first and foremost patient observers. This attitude of waiting to see what the animals do of their own accord, instead of encouraging a particular kind of behavior for experimental purposes, also characterizes our research in Arnhem.

Perception

Everyone can look, but actually perceiving is something that has to be learned. This is a constantly recurring problem when new students arrive. For the first few weeks they "see" nothing at all. When I explain to them at the end of an aggressive incident in the colony that "Yeroen rushed up to Mama and slapped her, whereupon Gorilla and Mama joined forces and pursued Yeroen, who sought refuge with Nikkie," they look at me as if I am crazy. Whereas to me this is a superficial summary of a fairly simple interaction (only four chimpanzees were involved), the students have only seen a few black beasts chaotically charging around uttering ear-piercing screams. They probably will have missed the hard slap.

At such times I have to remember that I too went through a long period when I found myself wondering at the apparent lack of structure in these episodes, whereas the real problem was not the lack of structure but my own lack of perception. It is necessary to be completely familiar with the many individuals, their respective friendships and rivalries, all their gestures, characteristic sounds, facial expressions, and other kinds of behavior. Only then do the wild scenes we see actually begin to make sense.

Initially we only see what we recognize. Someone who knows nothing about chess and who watches a game between two players will not be aware of the tension on the board. Even if the watcher stays for an hour, he or she will still have great difficulty in accurately reproducing the state of play on another board. A grand master, on the other hand, would grasp and memorize the position of every piece in one concentrated glance of a few seconds. This is not a difference of memory, but of perception. Whereas to the uninitiated the positions of the chess pieces are unrelated, the initiated attach great significance to them and see how they threaten and cover each other. It is easier to remember something with a structure than a chaotic jumble.

This is the synthesizing principle of the so-called *Gestalt perception:* the whole, or Gestalt, is more than the sum of its parts. Learning to perceive is learning to recognize the patterns in which the components regularly occur. Once we are familiar with the patterns of interaction between chess pieces or chimpanzees, they seem so striking and obvious that it is difficult to imagine how other people can get bogged down in all kinds of detail and miss the essential logic of the maneuvers.

Communication Signals

Each facial expression indicates a particular mood. The difference between a playful and an anxious mood, for example, can be deduced from the extent to which the teeth are bared. When chimpanzees are frightened or distressed, they bare their teeth much further than when they put on the so-called playface. To ordinary onlookers the wide-mouthed

Chimpanzees bare their teeth when they are frightened, uncertain, or uncomfortable: left, *Roosje reacts by screaming when her security blanket is taken away;* right, *Yeroen grins and yelps while avoiding an intimidation display by Nikkie.*

The playface (here Tarzan and Jakie) can be seen during wrestling and tickling games. This may be accompanied by a panting sound that is strongly reminiscent of suppressed laughter.

Screaming is the loudest form of vocalization and expresses frightened protest. Here, an adult male, Luit, is screaming after having been attacked by a group of females.

Screaming, Jakie holds out a hand in a begging gesture to another chimpanzee who has stolen his berries. He wants them back.

expression resembles a happy grin, but you can be sure that as far as the chimpanzee is concerned there is nothing to laugh about. This grin is seen on an infant who is momentarily left alone by its mother, or on older apes who get into conflict with high-ranking members of the group (who themselves seldom show their teeth).

Vocalizations often accompany these fearful facial expressions. The loudest of them is screaming. During the period when Yeroen, the oldest male, was being dethroned, his screaming could be heard throughout the zoo. I always ate my lunch while walking through the park, and in that period I would frequently hear Yeroen in the distance engaged yet again with his challenger. I would quickly gulp down my sandwich and hasten over to the enclosure to watch the spectacular scenes.

This screaming, which can be called a form of frightened protest, often changes to a yelp, a softer sound more like a disappointed whine. Chimpanzees also communicate by barking, grunting, whimpering, and hooting. The best way to learn to recognize the individual sounds is to make a tape recording of all of them and to play it over and over again until the difference becomes apparent. It is just like music from a strange culture; the melodies only emerge after frequent, repeated listening.

When familiarizing ourselves with chimpanzee communication we have the additional problem of the great differences between individuals. Each ape develops a number of special signals. Dandy, for example, has his own gesture to invite others to come and groom him: he holds his left upper arm with his right hand. It goes unnoticed when he is sitting, but when he hobbles towards a potential grooming partner on one arm, which he is holding with the other, and two legs, one might almost think he were crippled. Another strictly personal signal is the way Mama shakes her head to say no, and which really does look as if she means no. An example: Mama holds out her hand in begging fashion to Gorilla, whereupon another female comes and sits between Mama and Gorilla. Mama then shakes her head firmly from side to side. The other female's reaction is to withdraw hesitantly, after which Mama invitingly stretches out her arm to Gorilla once again. Gorilla comes and sits down beside her, and they begin to groom each other.

We call the gesture with the extended arm and open palm "holding out a hand." It is the most common hand gesture in the colony. Its significance, as with so many chimpanzee signals, depends on the context in which it is used. The apes use it to beg for food, for bodily contact, or even for support during a fight. When two apes confront each other aggressively, one of them may hold out his hand to a third ape. This gesture of invitation plays an important role in the formation of aggressive alliances, or coalitions: the political instrument par excellence.

All the (more than a hundred) behavior patterns that are regularly observed in our colony have also been observed among chimpanzees in their natural habitat. The playface, the grin, and the begging gesture are not imitations of human behavior, but natural forms of nonverbal communication that humans and chimpanzees have in common. Some unusual signals, such as the way Mama shakes her head to say no, may very well be the result of human influence. But even this very special signal was observed by Adriaan Kortlandt among wild chimpanzees. By and large the communication between the apes in the Arnhem colony is no different from the communication between their wild counterparts.

Side-Directed Behavior

Imagine a situation where one of the adult males displays at his rival. He looks all puffed up because his hair is on end, he is hooting, the upper part of his body is swaying from side to side, and he has a stone in his hand. Inexperienced observers would not notice the stone because

One of the most expressive ways in which chimpanzees communicate is by making their hair stand on end. Here Nikkie makes himself look as large as possible while displaying at Yeroen.

Chimpanzees sometimes hoot as a form of contact over longer distances. Luit (standing) and Yeroen answer Nikkie's hoot. Nikkie is displaying 60 meters away.

they would be so fascinated by the striking bluff display. They may be so fascinated that they would also fail to see the manipulative action of one of the adult females. She calmly walks up to the displaying male, loosens his fingers from around the stone, and walks away with it. It took me many weeks of observation before I realized what was happening. The entry in my diary for that day has a bold exclamation mark against it because at the time I was convinced that I had made the discovery of the century. But once I became familiar with the pattern I realized that it was not an unusual occurrence at all. It sometimes happens several times a day. We call it *confiscation.* In such a situation the male has never been known to react aggressively towards the female. He does sometimes try to pull his hand away and, if this fails, he may look for another stone or stick. Then he continues his bluff display. But even this second weapon can be confiscated: on one occasion a female confiscated no fewer than six objects from a single male.

The young chimps learn from watching the interactions between their elders:
top, *with his hair on end Fons follows Yeroen who is chasing away a challenger,*
screaming as he does so; bottom, *from the safety of her mother's lap Roosje*
observes a quarrel between two chimp children.

Patterns of social interaction are even more difficult to learn to recognize than communication signals such as hand gestures and vocalizations. Confiscation is one example, but there are many others. It is above all the aggressive interactions that cause problems. Conflicts may remain confined to two chimpanzees, but often other members of the group will interfere, so that eventually three or as many as fifteen individuals will be threatening and chasing one another simultaneously. On such occasions the chimpanzees perform highly complex patterns, which are accompanied by a great deal of noise.

In order to understand what is happening, the first distinction to be made is between behavior toward opponents and behavior toward companions or outsiders. The latter is called "side-directed behavior" and may take the following forms:

Seeking Refuge and Reassurance

This is the most common form. Young apes often react in this way when they have lost a fight with a peer or are threatened by an adult. On such occasions the young ape runs screaming to his mother and buries himself in her arms. Among adult apes things are done differently. A female who is being threatened may run to the most dominant male and sit down beside or behind him, whereupon the attacker will not dare to proceed.

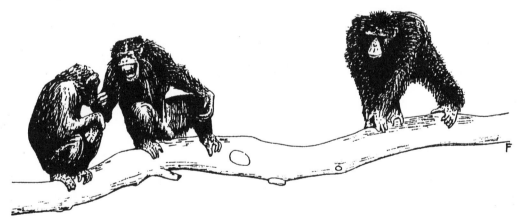

The screaming male in the center is simultaneously interacting with two other chimpanzees. On the right his bluffing opponent approaches. Before side-stepping his opponent he seeks reassurance from the female on the left by putting a finger in her mouth. This is an example of side-directed communication. (Left to right, Mama, Luit, and Nikkie.)

Luit kissing Wouter (photo: Ronald Noë).

An excited or frightened chimpanzee clearly has an urgent need for physical contact with others. It seems to be the only thing that has a rapid calming effect. The need to seek reassurance in the face of aggression assumes such proportions that the opponents seem to forget each other in the process. For example, a screaming, yelping adult male runs across the enclosure to several apes, young and old, seeking to touch, kiss, or embrace them. At that point the situation looks friendly enough, but it began with a lengthy, challenging bluff display by another male. This rival's hair is still on end, and he will shortly start threatening his screaming opponent again.

Recruitment of Support

Begging for support is done, as I described earlier, by holding out a hand. That this is actually intended as an appeal becomes obvious as soon as the other chimpanzee stands up and goes over with the victim to the enemy, because then the "beggar's" attitude changes dramatically. He is no longer the yelping, threatened creature holding out his hand; barking and screaming aggressively he charges at his opponent, always looking round to make sure his supporter is still backing him up. If the supporter appears to hesitate, the begging procedure may begin anew.

Instigation

Here communication takes place simultaneously in two directions. In most cases it involves females who recruit a male to attack another female. The threatened female challenges her opponent with a high-pitched, indignant bark, at the same time kissing and making a fuss of the male. Sometimes she points at her opponent. This is an unusual hand gesture. Chimpanzees do not point with a finger but with their whole hand. The few occasions on which I have seen them actually point have been when the situation was confused; for example, when the third party had been lying asleep or had not been involved in the conflict from the start. On such occasions the aggressor would indicate her opponent by pointing her out.

A characteristic feature of instigation is that females who have done the instigating do not join in when the male undertakes some action. They leave him to do the job on his own.

Reconciliations

Traditionally, aggression has been viewed as an uncontrollable instinct resulting in the dispersal of individuals. This spacing function is obvious enough in the territorial species that the early ethologists studied. However, how could it ever apply to social animals? Would not a group dissolve very quickly if every squabble drove its members apart? How do animals manage to compete over food and mates while maintaining cohesive groups? When we collected data on what happens after conflicts in the chimpanzee colony, we found former opponents to be attracted to each other like magnets! Much of the time, we registered contact rather than avoidance in the aftermath of fights.

In the months before my first student, Angeline van Roosmalen, arrived, I had gradually begun to realize the existence of the phenomenon of reconciliation among chimpanzees. Sometimes the maneuver is fairly obvious. Within a minute of a fight having ended the two former opponents may rush towards each other, kiss, and embrace long and fervently and then proceed to groom each other. But sometimes this kind of emotional contact takes place hours after a conflict. When I observed very carefully, I saw that the tension and hesitancy remained as long as the opponents had not reconciled their differences. Then suddenly the ice would break, and one of the chimpanzees would approach the other.

Mama mediates in a squabble between Nikkie (right) *and the screaming Fons* (left). *She "greets" Nikkie; in a moment she will embrace and kiss him. Only after Nikkie has been mollified by her in this way will Fons dare to become reconciled with Nikkie.*

Angeline was able to show that contacts between opponents after a conflict are much more intense than contacts in other situations, kissing being the most characteristic feature. The obvious word to describe this phenomenon is *reconciliation,* but I have heard people object to it on the grounds that by choosing such terms chimpanzees are unnecessarily humanized. Why not call it something neutral like "first post-conflict contact," because after all that is what it is? Out of the same desire for objectivity, kissing could be called "mouth-mouth contact," embrace "arms-around-shoulder," face "snout," and hand "front paw." I am inclined to take the motives put forward in favor of dehumanized terminology with a pinch of salt. Is it not an attempt to veil in words the mirror that chimpanzees hold up to us? Might we not be sticking our heads in the sand to preserve our sense of dignity?

People who work with chimpanzees know from their own experience just how strong the need for reconciliation is. Probably no other animal species demonstrates this need so forcefully, and it takes some getting used to. Yvonne van Koekenberg has described her reaction to her first experience of this phenomenon.

Yvonne had a young chimpanzee called Choco staying with her for a while. Choco was becoming more and more mischievous, and it was time she was checked. One day when Choco had taken the phone off the hook for the nth time, Yvonne gave her a good scolding while at the same time gripping her arm unusually tightly. The scolding seemed to have the desired effect on Choco, so Yvonne sat down on the sofa and started to read a book. She had forgotten the whole incident when suddenly Choco leapt on to her lap, threw her arms around Yvonne's neck and gave her a typical chimpanzee kiss (with open mouth) smack on the lips. This was so completely different from Choco's usual behavior that it must have been connected with the scolding. Choco's embrace not only moved Yvonne, it also gave her a deep emotional shock. She realized that she had never expected such behavior from an animal; she had completely misjudged the intensity of Choco's feelings.

Since our initial observations in Arnhem, reconciliation has become a popular topic of research. The phenomenon has been found in a wide range of primates, both in captivity and in the wild. Currently, there is no doubt that primates are capable of reconciling; the question rather is under which circumstances they do so. The most influential idea is that reconciliation serves the repair of valuable relationships. This would explain why in species after species we see reconciliations chiefly between individuals with close ties and cooperative partnerships.

In *Peacemaking among Primates* I reviewed the evidence, including the discovery that what chimpanzees do with kissing and embracing, the chimpanzee's closest relative, the bonobo, does with sex. After a fight between two bonobos—regardless of whether they are of the opposite or the same sex—copulation, pseudo-copulation, or mutual genital contact are common rituals. The point of these contacts is the same as it is for chimpanzees: both species share the need to resolve conflicts.

Coalitions

When two apes come to blows or threaten each other a third ape may decide to enter the fray and side with one of them. The result is a coalition of two against one. In many cases the conflict extends still further, and larger coalitions are formed. Because everything happens so

Cooperation is not only demonstrated in coalitions: Tepel helps Tarzan out of a tree.

quickly, we might imagine that chimpanzees are carried away by the aggression of others and that they join in blindly. Nothing is further from the truth. Chimpanzees never make an uncalculated move.

In order to prove this we have to check repeatedly just what each individual does in the mêlée. Does he or she intervene in an unpredictable manner, or does he or she systematically support certain individuals? This calls for very careful observation; collecting information about coalitions requires patience, patience, and still more patience. Sometimes it is possible to wait an entire day and not witness one instance.

The average, however, is between five to six coalitions a day, and by observing very closely we as a team are able to record a total of 1,000–1,500 coalitions a year. These are recorded in long lists, in the form of "C supports A against B." Analysis of these lists confirms that chimpanzees act selectively when intervening in a conflict between other members of the group. All the group members have their own personal likes and dislikes that dictate how they act. The choices they make are biased choices, which generally remain constant over the years.

This is not to say that relationships in the group do not change; indeed, this is the most fascinating aspect of chimpanzee coalitions. Why should C, who has supported A against B for years, gradually begin to support B against A? Where is the strongest element of change, in the A—B, B—C, or A—C relationship? The problem is complex because it concerns a three-cornered relationship. And the ABC combination is only one of the many thousands of triadic relationships that exist within the group. Studying coalitions confronts us with the full extent of the "third dimension" of group life.

Two primatologists, Irven DeVore and the late Ronald Hall, published the first full-scale study on this subject in 1965. They had studied the behavior of free-ranging baboons in Kenya. The status of an adult male baboon depends upon both his individual fighting ability and joint action. The whole baboon troop is led by two or three adult males jointly, who form the so-called "central hierarchy." Individually, without each other's support, none of them need carry much weight. Some males outside this central coalition show no fear at all when confronted by only one of the central males. In order to keep their rivals under control the central hierarchy has to form a common front.

A few years ago Ron Nadler described another, marvellous example of how the top position in a group may depend on aggressive cooperation. A group of gorillas was formed at the Yerkes Primate Center in Atlanta. The group comprised four adult females plus Calabar, a large, impressive male, and Rann, a much smaller adult male. Everyone expected Calabar to become the leader of the group, but the females favored Rann. Although both males had cohabited quite peacefully for weeks, their introduction into the group of females resulted in chest-beating, charging displays, and serious fights. Nadler described the final fight, in which Calabar was injured and had to be removed from the group: "It was not clear which male made the first move but, once they became locked together in combat, the females joined the struggle. Two jumped on Calabar's back, one grabbed at a leg, and they all began biting him. They struggled furiously, but briefly, all parting in a matter of seconds."

The fact that it was the females who helped their chosen male into the position of leader is not even the most striking aspect of this incident. The most surprising thing is that Rann was able to *coerce* the females into supporting him. This became evident from the maneuvers that preceded the fight: "Wherever Rann stalked Calabar, the females would soon follow. Whenever Calabar stopped, they, together with Rann, formed a semicircle around the impressive male. In fact, when, at one point, one of the females started to leave the encircling group, Rann charged at her and chased her back into position." This was Rann's way of prohibiting desertion. But why did the females obey a male who was at the same time so dependent on them? After all, his fate was in their hands. Perhaps gorilla politics are as refined, intricate, and puzzling as those of chimpanzees.

We know enough about coalitions among chimpanzees in their natural habitat to conclude that they are very important in determining relationships between adult males and establishing their respective dominance. This is something which is repeatedly emphasized in the publications on the Gombe Stream community. We have an almost complete picture of the gradual development of a coalition there, between the brothers Faben and Figan. If we compare these processes with the processes that take place in the Arnhem colony, there are no obvious, fundamental differences. The only difference is that in Arnhem we were able to study the processes in infinitely greater detail.[4]

Safe Interpretations

How do we recognize an animal's mood? When a dog puts his tail between his legs, we say he is frightened. This is because we have learned that a dog that acts in this way generally tends to run away. The act of running away is not nearly as difficult to understand as the act of "tail between legs." From the connection that exists between the two we deduce that, if the one act expresses fear, then the other act will express fear too. Analogously we know only too well what a dog means when he growls deeply and his hackles rise. These are associations which we have learned to make unconsciously, and that is why we often ascribe them to intuition; we say that we "intuitively know" the dog's various moods.

Intuition is valuable, but scientists are only fully satisfied when they know what lies behind it. It is not necessary to depend on this suggestive indicator all the time. The unconscious method we have learned to use to interpret and understand the signals of a dog can be translated into an effective, scientific tool if it is applied consciously and systematically. Jan van Hooff has applied this method to the piles of notes he made on social behavior in the Holloman colony. His notes indicated the order in which the chimpanzees had displayed all kinds of behavior patterns. He used a computer to sort out which patterns had frequently occurred simultaneously or in rapid succession. The result was a number of sets of inter-connected behavior patterns. A set with patterns such as fleeing, evading, and parrying was designated "submissive"; a set with patterns such as attacking, biting, and trampling was designated "aggressive," and so on. From this it was possible to deduce the less obvious interpretations. Barking, for example, was found to belong to the aggressive set (attack) and screaming and yelping to the submissive set (flight).

The computer only provides associations between behavior patterns; it cannot indicate what lies behind them. That is why van Hooff cautiously calls the sets "behavior systems" instead of emotions or motivations. For convenience's sake I do not intend to imitate this caution. When I say that a chimpanzee "pants in a friendly fashion" to another chimpanzee, I mean that he is breathing audibly and that this panting, according to van Hooff's analysis, can be called "affinitive" behavior. This behavior set can be so described because it contains several obviously affinitive forms of contact, such as embracing, kissing, and social grooming.

Daring Interpretations

Diametrically opposed to the concept of instinctive and impulsive animal behavior is the concept of conscious, premeditated action. There are, of course, many animals that are probably totally unaware of the consequences of their social behavior. Does a male cricket, for example, know that his chirping attracts females? And yet that is the function of his signal. Higher animals, however, do seem to know the effects of their signals. Great apes, in particular, behave so flexibly that we get the impression that they know exactly how others will react, and what they can achieve as a result. Their communication looks very much like intelligent social manipulation, as if they have learned to use their signals as instruments to influence others.

Example 1

On a hot day two mothers, Jimmie and Tepel, are sitting in the shadow of an oak tree while their two children play in the sand at their feet (playfaces, wrestling, throwing sand). Between the two mothers the oldest female, Mama, lies asleep. Suddenly the children start screaming, hitting, and pulling each other's hair. Jimmie admonishes them with a soft, threatening grunt, and Tepel anxiously shifts her position. The children go on quarrelling, and eventually Tepel wakes Mama by poking her in the ribs several times. As Mama gets up Tepel points to the two quarrelling children. As soon as Mama takes one threatening step forward, waves her arm in the air, and barks loudly the children stop quarrelling. Mama then lies down again and continues her siesta.

INTERPRETATION. In order to understand this interpretation fully, it is important to know two things: first, that Mama is the highest-ranking female and is greatly respected; and second, that conflicts between children regularly engender such tension between their mothers that they too come to blows. This tension is probably caused by the fact that each mother wishes to help her own child and to prevent the other from interfering in the quarrel. In the case of the example above, when the children's game turned to fighting, both mothers found themselves in a painful situation. Tepel solved the problem by activating a dominant third party, Mama, and pointing out the problem. Mama obviously realized at a glance that she was expected to act as arbitrator.

Example 2

Yeroen hurts his hand during a fight with Nikkie. Although it is not a deep wound, we originally think that it is troubling him quite a bit, because he is limping. The next day a student, Dirk Fokkema, reports that in his opinion Yeroen limps only when Nikkie is in the vicinity. I know Dirk as a keen observer, but this time I find it hard to believe him. We go to watch, and it turns out that he is indeed right: Yeroen walks past the sitting Nikkie from a point in front of him to a point behind him and the whole time Yeroen is in Nikkie's field of vision he hobbles piti-

The advanced mental processes of chimpanzees were first demonstrated by Köhler's famous tool experiments. Chimpanzees are spontaneous in their use of instruments. Here Amber has seen a piece of apple peel floating on the water. She is trying to reach it with a stick. Zwart (left) and Franje are curious to see if she will succeed.

fully, but once he has passed Nikkie his behavior changes and he walks normally again. For nearly a week Yeroen's movement is affected in this way whenever he knows Nikkie can see him.

INTERPRETATION. Yeroen was playacting. He wanted to make Nikkie believe that he had been badly hurt in their fight. The fact that Yeroen acted in an exaggeratedly pitiful way only when he was in Nikkie's field of vision suggests that he knew that his signals would only have an effect if they were seen; Yeroen kept an eye on Nikkie to see whether he was being watched. He may have learned from incidents in the past in which he had been seriously wounded that his rival was less hard on him during periods when he was (of necessity) limping.

Example 3

Wouter, a young male chimpanzee of almost three, gets into a quarrel with Amber and screams at the top of his voice. At the same time he advances aggressively towards Amber. His mother, Tepel, goes over to him and quickly places her hand over her son's mouth, smothering his screams. Wouter calms down and the quarrel is over.

INTERPRETATION. Noisy conflicts attract attention. If they last too long, one of the adult males will come over and put an end to them. When a bluffing male approaches, Wouter will automatically seek refuge near his mother. This means that she runs the risk of receiving the punishment meant for her son. Tepel wanted to avoid this risk by shutting up Wouter before things went too far.

This is not the only known instance of enforced silence. I have also seen a mother place a finger over the small mouth of her baby when the latter started barking aggressively at a dominant group member from the safety of her lap. Once again this was probably due to the mother's reluctance to get drawn into difficulties because of a social faux pas committed by her child.

Example 4

Dandy is the youngest and lowest ranking of the four grown males. The other three, and in particular the alpha male, do not tolerate any sexual intercourse between Dandy and the adult females. Nevertheless every now and again he does succeed in mating with them, after having made a "date." When this happens the female and Dandy pretend to be walking in the same direction by chance, and if all goes well they meet behind a few tree trunks. These "dates" take place after the exchange of a few glances and in some cases a brief nudge.

This kind of furtive mating is frequently associated with signal suppression and concealment. I can remember the first time I noticed it very vividly indeed, because it was such a comical sight. Dandy and a female were courting each other surreptitiously. Dandy began to make advances to the female, while at the same time restlessly looking around to see if any of the other males were watching. Male chimpanzees start their advances by sitting with their legs wide apart revealing their erection. Precisely at the point when Dandy was exhibiting his sexual urge in this way, Luit, one of the older males, unexpectedly came around the corner. Dandy immediately dropped his hands over his penis concealing it from view.

On another occasion Luit was making advances to a female while Nikkie, the alpha male, was lying in the grass about 50 meters away. When Nikkie looked up and got to his feet, Luit slowly shifted a few paces away from the female and sat down, once again with his back to Nikkie. Nikkie slowly moved towards Luit, picking up a heavy stone on his way. His hair was standing slightly on end. Now and then Luit looked round to watch Nikkie's progress and then he looked back at his own penis, which was gradually losing its erection. Only when his penis was no longer visible did Luit turn around and walk towards Nikkie. He briefly sniffed at the stone Nikkie was holding, then he wandered off leaving Nikkie with the female.

Females sometimes give away their clandestine mating sessions by emitting a special, high scream at the point of climax. As soon as the alpha male hears this he runs towards the hidden couple to interrupt them. An adolescent female, Oor, used to scream particularly loudly at the end of her matings. However, by the time she was almost adult she still screamed at the end of mating sessions with the alpha male, but hardly ever during her "dates." During a "date" she adopted the facial expression that goes with screaming (bared teeth, open mouth) and uttered a kind of noiseless scream (blowing from the back of the throat).

INTERPRETATION. In all these examples sexual signals are either concealed or suppressed. Oor's noiseless scream gives the impression of violent emotions that are controlled with only the greatest effort. The males are faced with the problem that the evidence of their sexual arousal cannot disappear on command, but they too have their solutions.

The audacity of Luit actually sniffing at the weapon Nikkie held in his hand only goes to show how sure he was that the alpha male would find no cause to proceed against him. This behavior is in marked contrast to an incident I once witnessed between two male macaques. The alpha male met another male several minutes after the latter had secretly mated. Alpha could not possibly have known anything about this, but the other male acted unnecessarily timidly and submissively. His behavior was so exaggerated that, if the alpha male had had a chimpanzee's social awareness, he would certainly have realized what the matter was.

Luit's behavior after his abortive adventure was very different. There was no trace of a "guilty conscience." Chimpanzees are masters of pretense and will seldom put an idea into the head of the unsuspecting.

Rational Behavior

Once we have witnessed a number of striking instances of social manipulation and recognized that chimpanzees are more than highly intelligent, we are forced to consider the nature of the extra faculty they have which most other species appear to lack: the ability to *think purposefully.*

When a rat is trained to push down a pedal to get food, it will use the pedal when it is hungry and it will stop as soon as it has had enough to eat. The rat acts in this way purely because it has discovered, more or less by accident, that by touching the pedal it causes food to be released and it has remembered this fact. Some goal-directed behavior occurs among chimpanzees, however, without there being any past proof of the effectiveness of the result. They seem to be able to devise effective, on-the-spot solutions, such as in example 1, when Tepel woke Mama and pointed to the two quarrelling children, or in example 3, where she effectively silenced her son. It is hard to see how Tepel could have accidentally discovered that these actions would extricate her from the tricky situations in which she found herself. Surely far more was needed here than just a good memory?

On the other hand, how can such solutions be separated from Tepel's social experience? She showed a striking ability to relate effectively a whole series of past experiences, including knowledge of children's quarrels, sleeping apes, Mama's position of authority, and the effect of a hand placed over a mouth. The extra faculty that makes chimpanzee behavior so flexible is their ability to *combine* separate bits of knowledge. Because their knowledge is not limited to familiar situations, they do not have to feel their way blindly when confronted with new problems. Chimpanzees use all their past experience in ever-changing practical applications.

The ability to combine past experiences in order to achieve a goal is best described as *reasoning* and *thought;* no better words exist. Instead of testing a particular course of action through actual trial and error, chimpanzees are able to weigh the consequences of a choice in their heads. The result is considered, rational behavior. Primates take such a mass of social information into account, and are so finely attuned to the moods and intentions of others, that it has been speculated that their high intelligence evolved in order to deal with an increasingly complex group life. This idea, known as the Social Intelligence Hypothesis, may also apply to the enormous brain expansion in our own lineage.[5]

In this view, technical inventiveness is a secondary development: the evolution of primate intelligence started with the need to outsmart others, to detect deceptive tactics, to reach mutually advantageous compromises, and to foster social ties that advance one's career. Chimpanzees clearly excel in this domain. Their technical skills are inferior to ours, but I would hesitate to make such a claim with regard to their social skills.

PERSONALITIES

CHIMPANZEES HAVE OUTSPOKEN PERSONALITIES. THEIR FACES ARE FULL of character, and you can distinguish them one from another just as easily as you can distinguish people. Also, their voices all sound different, so that years later I can tell them apart by ear alone. Each ape has his or her very own way of walking, lying down, and sitting. Even by the way they turn their heads or scratch their backs, I can recognize them. But when we speak of personality of course we think especially of the differences in the way in which they treat their groupmates. These differences can only be portrayed accurately by using the same adjectives as we use for our fellow humans. Therefore, terms such as *self-assured, happy, proud,* and *calculating* will be used in this chapter of first acquaintance with the individuals. These terms reflect my subjective impression of the apes. It is anthropomorphism in its purest form.

CHAPTER ONE

That chimpanzees are experienced in the first place as personalities is evident from the dreams of those of us who work with them. We dream about these apes as individuals, in the same way that other people dream about their fellow human beings as individuals. If a student were to say that he or she had dreamed of an ape I would be no less surprised than if someone claimed to have dreamed of a human.

I clearly remember the first dream I had about the chimpanzees. In it my preoccupation with the distance between them and me was apparent. During this dream the large door to their quarters was opened for me from the inside. The apes were pushing each other aside in order to get a good look at me. Yeroen, the oldest male, stepped forward and shook my hand. Rather impatiently he listened to my request to come in. He refused point blank. That was out of the question, he said, and besides, their society would not suit me: it was much too harsh for a human being.

Students of nonprimates used to criticize the habit of primatologists of giving each individual a name. They charged that naming led to an unnecessary humanization of the animals. The hidden implication was that attention to individual differences was not nearly as important as a search for species-typical behavior. Nowadays, of course, it is not just the primatologists who realize that animal behavior makes little sense unless one factors in the unique genetic make-up, life history, and social background of each individual. The first scientists to use indi-

Dandy (with Spin)

Four Males

Yeroen

Luit

Nikkie

Dandy

Female Subgroup "Mama"

Amber

Mama & Moniek

Gorilla & Roosje

Fons

Franje

Female Subgroup "Jimmie"

Jimmie & Jakie

Jonas

Krom

Spin

Female Subgroup "Tepel"

Tepel

Tarzan

Wouter

Puist

Three Girls

Zwart

Oor

Henny

vidual identification on a large scale were the Japanese primatologists, who started this practice in the 1950s. They used number codes, which may have made their observations sound more objective than those by Jane Goodall, who adopted names such as "Humphrey" and "Flo," but the principle remained the same. Every observer who has tried number codes reports that after a while the numbers start to sound like names, probably because we, humans, automatically think in terms of named personalities.

In 1979, when I began preparing for this book, the colony had twenty-three members. Seven of these apes, three females and four males, were particularly influential and they are described individually. The other sixteen were mostly females and infants belonging to three female sub-groups formed around the first mothers in the colony. The estimates of the apes' ages apply to 1979.

Big Mama

Our oldest chimpanzee is a female whom we estimate to be about forty. (The maximum recorded age for a chimpanzee in captivity is fifty-nine. In the wild they probably never survive for so long.) We call her Mama. There is great power in Mama's gaze. She looks at us in the en-quiring and all-comprehending manner of an old woman.

Mama enjoys enormous respect in the community. Her central posi-tion is comparable to that of a grandmother in a Spanish or Chinese family. When tensions in the group reach their peak, the combatants always turn to her—even the adult males. Many a time I have seen a major conflict between two males end up in her arms. Instead of resort-ing to physical violence at the climax of their confrontation, the rivals run to Mama, screaming loudly.

The most convincing demonstration of her conciliatory role came on an occasion when the whole group had turned on Nikkie. Only a few months previously, Nikkie had become the alpha male and his violent actions were still frequently not accepted. This time, all of the apes, in-cluding Mama, had given chase, screaming loudly and barking. In the end the usually so impressive Nikkie sat high up in a tree, alone, panic-stricken and screaming. Every line of escape was cut off. Each time he wanted to come down, some of the others chased him back. After about a quarter of an hour, the situation changed. Mama slowly climbed into

Mama fulfils a central role in the group. Apart from her stabilizing and conciliatory influence, she is also the leader of collective female power. None of the males can ignore her.

the tree, touched Nikkie and kissed him. Then she climbed down again with Nikkie following close at heel. Now that Mama was bringing him with her, nobody resisted any more. Nikkie, obviously still nervous, made up with his opponents.

Mama is a lady of considerable size. She is exceptionally broad and strongly built. She walks slowly, and climbing is quite an effort. She sometimes pulls a face which makes us think that perhaps the burden on her joints causes her pain. When the colony was first established she was much faster; necessarily so because she was the group leader, dominating not only the adult females but the adult males as well.

The adult males were added to the colony rather late. Mama had already been in the leading position for about eighteen months when suddenly, on 5 November 1973, three adult males appeared on the scene. They laid no claim to her power; on the contrary, they had enough difficulty in keeping the biting, tugging, and hitting womenfolk at bay.

It was winter and consequently the chimpanzees lived in the large indoor hall. Each morning the three new males were let out first. One of them, Yeroen, always ran straight to the large drums, hooting and with his hair on end. He was followed closely by the other two, both of whom screamed and kept looking over their shoulders fearfully. The three males stayed close together and kept their eyes on the corridor out of which the females would appear. The males would then stand on the tallest drum, one on top of the other, while the females attacked them from beneath. All this went on under the command of Mama and her

*Three large males huddle fearfully
together on the highest drum
while the females crowd around
screaming aggressively and
pulling at their feet and hair.
A female is displaying in the
foreground.*

friend, Gorilla. They bit the males' feet and pulled their hair. The victims defended themselves as best they could, but this only aggravated the aggressive display of the others. Their fear of the infuriated females was evident in their intense screaming, diarrhea, and vomiting.

After a number of days some chimpanzees started contacting the males cautiously. Also Gorilla, Mama's mainstay, behaved in an increasingly friendly way towards the new arrivals. She showed a definite preference for Yeroen. This preference remained and was to play an important part in the developments during subsequent years. Even so, despite the hesitant advances between both parties, violent conflicts burst out from time to time. Mama saw her position threatened and was not in the least inclined to accept the males.

These conditions came to an end through human interference. Two weeks after the introduction of the males, Mama and Gorilla were removed from the group. In the following months Yeroen was able to impress the remaining chimpanzees to such an extent that they submitted to him. Yeroen achieved this mainly by so-called drumming concerts: lengthy, rhythmic stamping on the large, hollow, metal drums. While he was doing this the whole building boomed with the sound of his drumming. He would conclude his drum solos by suddenly jumping in among the others, his hair on end, and performing a wild charge. Whoever failed to get out of his way fast enough would get a beating.

After three months it was clear that Yeroen was firmly in the saddle. Only now were Mama and Gorilla allowed back into the group. Naturally this was a most embarrassing confrontation! The student who witnessed and recorded all these spectacular scenes, Titia van Wulfften Palthe, wrote in her report:

> The entry of Mama and Gorilla into the hall caused enormous excitement among the chimpanzees, and was accompanied by a deafening din. The scene was similar to the one in the very beginning. The three adult males piled one on top of the other on the highest drum and were screaming incessantly. Again they had diarrhea for fear of the females. When some of it landed on Mama's leg she took her time to clean her coat thoroughly with a piece of frayed rope. Mama was extremely aggressive towards the three males and attacked them as much as possible.

But now there was one noticeable difference. Mama got no support from any of the other group members, not even from Gorilla, who had

Mama and Moniek.

already, ten minutes after entering, made friendly contact with Yeroen. The falling apart of the solidarity among the females meant that the fear of Mama diminished rapidly. A few weeks later she was dethroned. From that moment on, Yeroen was the boss.

The temporary removal of Mama and Gorilla from the group broke the strong alliance that formed the basis of their authority over the males. Even though this occurred well before I began my studies in Arnhem, I have been reproached for this intervention, especially by feminists. Why did we want to see the males in control? Was Mama not good enough?

There were a number of reasons for the interference. It is known that in the wild the adult males are dominant. Most probably our males would also have gained control without human intervention, but later and with more difficulties. Presumably they would not have succeeded during the winter, in the oppressive indoor hall. In the extensive open-air enclosure the males would have been able to keep their distance from Mama and the other females. Perhaps then they would have gained courage, and slowly, week by week, they might have given increasingly provocative bluff performances. Outdoors they would also have had the opportunity to isolate Mama from her supporters in order to fight with her separately. Adult males are stronger than females, and faster.

This more natural course of events was not awaited because there was a very urgent reason for wanting to curb Mama's power. Before the males were introduced, apes were being injured at the rate of almost one a week and having to be kept apart for a while until they had recovered. The injuries were mainly Mama's work. Not only did she bite frequently, but she drew blood and sometimes ripped her victim's skin. Although males are far from gentle, they seldom show such damaging forms of aggression. They seem to control their aggression better. Moreover, they prevent the escalation of conflicts between females by intervention in their fights. As far as the number of injuries in the group was concerned, Yeroen's rise to power was a relief. The lowest ranking individuals especially benefited from the changeover. Instead of Mama's fierce attacks, there was Yeroen, who could be harsh, but who never went too far.[6]

Over the years Mama changed considerably. During the first years of the colony when she held the supreme power she would perform intimidation displays just like a male. She would walk about stamping and with her hair on end. Her speciality was giving a tremendous kick against one of the metal doors. In doing this she would sway her broad body between her long arms like a swing. Resting her hands on the ground, she would fling her feet against the door, delivering an enormous blow. The noise was explosive.

I have seldom heard these blows. Mama had already been dethroned for two years when I came to work in Arnhem. At that time she was going through a transitional period. She showed less of the masculine bluffing behavior but did not as yet have much interest in offspring. That year she gave birth to a baby but did not want to care for it herself. She kept trying to give it away to her friend Gorilla. In the end we had to take her infant away from her and bottle-feed it ourselves. This was the fate of many of the young chimpanzees born during the colony's first few years.

Mama did accept her next child, who was born two years later. It seems as though she only reconciled herself to her new position in the group from that moment. She became noticeably more relaxed and tolerant. Her daughter, Moniek, now lives like a princess. Mama is very tender and protective. Every member of the group realizes that the old female's anger will flare up to its former hurricane force if but a hair on her daughter's body is harmed. This way, Moniek inherits some of the enormous respect that her mother enjoys in the colony.

Yeroen and Luit

The two oldest males in the group, Yeroen and Luit, have known each other for a long time. They both came from the zoo in Copenhagen, and it is likely that before their introduction into our colony they had spent years together in the same cage. Right from the start Yeroen dominated Luit. He is probably several years older. We judge Yeroen to be about thirty and Luit twenty-five.

Luit has a playful, almost mischievous character. He radiates youthful vigor, while Yeroen makes a more staid impression. Yeroen's beard is greyer, and he walks and climbs less smoothly than Luit. All this is reason to judge Yeroen to be the older of the two, but the most important reason is his decreasing stamina. His bluff displays usually are not very lengthy. He can be hugely impressive but tires quickly. Sometimes, after a bout of displaying, he sits with eyes shut, panting heavily. If for some reason he continues with his intimidation display, he may slip or

Yeroen, the old fox.

stumble, or miss his grasp when jumping from one branch to another. These signs of fatigue do not escape the notice of his opponents. This was apparent during the period in which Luit was setting himself up as Yeroen's rival. During mutual bluff displays, Luit would redouble his efforts when he saw that Yeroen was tiring.

Because they share the same background, we could call Yeroen and Luit old comrades. However, the undeniable bond between them is often obscured by disagreement. In colony life they have become opposites. At best we might say they are rivalling friends. Their not seeing eye to eye is, in fact, somewhat surprising. I had always imagined that they would place themselves at the top of the group together. Perhaps it is just as well that this did not happen, because then the developments in the group might not have been nearly as interesting.

Luit

Because I have known both males in different roles over the years, I can judge their characters. This would otherwise not have been an easy matter. If, for instance, we have known a male only in the role of alpha male, we probably think that he is very self-confident. That need not be the case at all. As soon as his position is seriously threatened, the self-confidence may vanish.

Yeroen is calculating by nature. In an almost nervous way he keeps a close watch on his interests. No one else is considered when he is pursuing his goal. He is a real go-getter, as later episodes will demonstrate.

Yeroen not only has his increasing age and diminishing endurance to contend with—he also has a serious physical handicap. When he has an erection, his penis catches on a subcutaneous fold so that it does not protrude beyond its shaft. He has a normal mating drive and regularly mounts females but cannot impregnate them. He has been operated on twice, without much effect.

Luit is a much more sociable individual than Yeroen. He has an open and friendly character and sets great store by company. He practically always seems to be in a good mood and creates a "reliable" impression. Some students who know the chimpanzees well have told me independently of each other that Yeroen gives the impression that "he will cheat you before your very eyes," while Luit seems to be "someone you can depend on." Luit is proud and aware of his strength. Whenever he gives an intimidation display, it is always a beautiful sight, full of rhythm and vigor. No other chimpanzee is so impressive and yet so elegant at the same time.

Puist

Puist is an adult female with an exceptionally heavy build and she walks and stands in a burly manner. Chimpanzee experts who see her from the front are often surprised afterward to discover from her backview that she is a female. Sexually she is aberrant as well: she refuses to mate. Because she never conceives she sports a genital swelling every month, year in year out. (Just as in humans, in chimpanzees the menstrual cycle is disturbed during pregnancy and the nursing period.) Consequently she is regularly attractive to the males, but it is hands off.

Yet Puist is far from indifferent to sex. In the first place she masturbates. Although self-gratification is a notorious and frequently occur-

Puist, "the Madam"

ring phenomenon among apes in captivity, in our group Puist is the only one with this habit. Curiously enough she only does it when she is not in the "pink period." She makes rapid finger movements through her vulva for about a minute. We cannot tell anything from her face, but it must have pleasant effects, otherwise why would she do it?

In the second place she behaves in a lesbian fashion now and then. When another adult female shows a genital swelling, Puist may invite her to have sexual intercourse. Sometimes the female accepts and then

Puist mounts her briefly, thrusting in the same way as males do when mating.

Her interest in other females goes even further. When the adult males collect around a sexually attractive female, Puist is often to be found with them. At such moments there is a tense, competitive atmosphere among the males. Every now and again one of them gets ready to mate with the female, and at times it seems as if, like the males, Puist also has a say in whether intercourse is tolerated. Sometimes she joins forces with males who try to prevent this contact by aggression. On the other hand, if the female refuses the advances she can count on fierce support from Puist if the male tries to insist. Due to this unique regulatory role in sexual intercourse we sometimes jokingly call Puist "the Madam."

Everyone differs in their opinion as to which chimpanzee they find most congenial. As far as the least likeable group member is concerned, on the other hand, there is surprising consensus: absolutely everyone names Puist. She is even compared to a witch. She gives the impression of being two-faced and mean. Not only is Puist often in the company of the adult males but she is also "in league" with them. Except in the sexual context, she is generally not prepared to stand up for other females. While the other females help one another against male aggression, Puist actually joins the other party. When a male attacks a female Puist will sometimes charge up to the victim and bite or hit her. Also, she can successfully set the males on other females. It is not surprising that subordinate females are terrified of her.

Besides this malevolence, Puist has another trait, which we might call deceitful or mendacious. If Puist is unable to get ahold of her opponent during a fight, we may see her walk slowly up to her and then attack unexpectedly. She may also invite her opponent to reconciliation in the customary way. She holds out her hand and when the other hesitantly puts her hand in Puist's, she suddenly grabs hold of her. This has been seen repeatedly and creates the impression of a deliberate attempt to feign good intentions in order to square accounts. Whether we regard it as deceit or not, the result is that Puist is unpredictable. Low-ranking apes hesitate when she approaches: they mistrust her.

Except for a special bond with one of the females, Tepel, Puist falls outside the central group of females. She is the only adult group member who on average spends more time with adults of the opposite sex than with her own sex. She thus occupies an intermediate position between the male and the female spheres of colony life. Given that these spheres are growing apart more and more over the years, Puist may in time form an important binding element between them. Interestingly enough, in the wild Gombe community there was also a very large, masculine-looking female, Gigi. She was sterile and associated strongly with the males. The big difference between Gigi and Puist is that Gigi did not reject sexual intercourse with the males.[7]

Gorilla

Gorilla is a female chimpanzee with a face as black and a back as straight as a gorilla. Together with Mama and Puist she is one of the most influential females in the colony. But while the other two are both very powerfully built, Gorilla is slight and slim. In contrast to her delicate figure, she has a much fiercer character. She "knows what she wants." Her face has an aspect of resolution, and in everything she does she is determined. Presumably Gorilla's high social rank is derived from her firm bond with Mama. Together with Mama and another female, Franje, she came from Leipzig Zoo. Mama and Gorilla have supported each other right from the start. Not only do they often act together against attackers, they also seek comfort and reassurance from each other. When one of them has been involved in a painful conflict, she goes to the other to be embraced. They then literally scream in each other's arms. Sometimes this contact seems to give them instant courage, and both go after the opponent fiercely. Then not a single ape dares to stay put, not even the males.

Gorilla is fond of young chimpanzees, and she has given care and protection to Mama's and Franje's children, Moniek and Fons. For years she never got beyond being an "aunt" because the children to whom she gave birth herself all died within a few weeks. This certainly was not due to her way of handling her offspring. Presumably she produced too little milk.

This disappointing situation came to an end in 1979 as a result of a unique experiment which received wide publicity. In short, we taught Gorilla to bottle-feed a baby chimpanzee. This baby, called Roosje, was not her own child. After Roosje had been in human care for ten weeks, Gorilla adopted her. From that moment on Roosje clung to her foster mother and was entirely dependent on her. Gorilla was extremely care-

Gorilla

ful with her and not only fed her by bottle but, after about a week, started producing milk herself. Presumably her nipples were stimulated by Roosje's sucking. After a while Roosje obtained more than half her daily requirements from Gorilla and the rest from the bottle.

When the keeper, Monika ten Tuynte, and I started this experiment we encountered two immediate difficulties. The first was to be expected. Gorilla, who likes milk herself, tried to empty Roosje's bottle. We stopped her doing this by becoming angry and grumbling at her. The second difficulty was unexpected—Gorilla was not very attentive during the instructions. Every day Monika sat in front of her night cage with Roosje and showed her how to bottle-feed a baby. We had hoped that Gorilla would start imitating her, but we were not so lucky. Gorilla did not even look at Monika; she continued to stare in the opposite

Roosje ("Little Rose") is the first animal in the world to be reared on the bottle by one of her own kind. Opposite page: *The feeding lessons took place while Gorilla was in the sleeping quarters* (top left). *Our main task was to keep Gorilla from consuming the milk herself. Reluctant to give up a good drink* (bottom), *she screams in response to our scolding. But then, the big moment arrives— Gorilla finally finds the baby in the straw of her cage* (top right). Above: *Roosje being bottle-fed.*

direction. This looking away did not stem from lack of interest, because she wanted to be as close to the baby as possible. This phenomenon is also seen when a female joins the group with a newborn baby. Some apes, especially the young females, hang around the baby all the time, but demonstratively avert their glance as soon as the mother looks at them. In this way they hide their interest—perhaps if they were too blatant about it they would irritate the mother. Gorilla behaved in exactly the same way toward Monika, and maybe this is why imitation barely played a role in the learning process.

Instead, we had to resort to a conditioning procedure. Step by step we taught Gorilla everything, rewarding her with delectable tidbits. It was weeks after this training, when she had already adopted the baby, that she began to show signs of comprehension. She started doing things that we had not taught her to do but that were perfectly sensible. For instance, if Roosje choked, Gorilla would quickly pull the nipple out of her mouth and only replace it after the baby had burped. After this we felt that we could leave the feeding entirely up to Gorilla.

Being with Gorilla in the colony, Roosje is enjoying a much more natural youth than humans could ever have offered her. Also, for Gorilla herself, the success of this adoption experiment is of major importance. Every time one of her own children died, she would go into a kind of depression. For weeks on end she would sit huddled in a corner without reacting at all to the goings-on about her. Sometimes she would start screaming and yelping of her own accord. This changed after she adopted Roosje. In the years that followed, Gorilla was able to bottle-feed her own newborns in the same way, even though there was less of a need for this than anticipated. Roosje, who also nursed naturally, apparently stimulated Gorilla's milk production to such a degree that her own subsequent offspring needed little supplemental feeding.

Nikkie and Dandy

We have now met nearly all the principal actors in the political drama. Mama, Yeroen, Luit, Puist, and Gorilla could quite easily have lived on for years in a stable group had it not been for Nikkie. Nikkie is the young hero of this tale—not a glorious hero, nor a tragic one, but the moving force behind all the developments. His boundless energy and boisterous, provocative behavior have had the effect of a catalyst. Bit by bit he has disrupted the structure of the group. On cold days Nikkie keeps the others warm by his constant activity, and on hot days he disturbs their sleep.

Nikkie has the appearance of a country bumpkin, with his bundle of muscles, broad head, and somewhat dopey expression. But appearances can be deceptive. He is extremely bright and is the fastest, most acrobatic ape in the whole group. His intimidation displays are characterized by spectacular leaps and somersaults. Before Nikkie came to us he had been appearing in the *Holiday on Ice Revue*. When he reached puberty his owners felt they had to get rid of him, presumably because of his nascent sexual interest. He was also rapidly growing too strong, and his canines were beginning to come through (the teeth of a full-grown male chimpanzee are as dangerous as those of a panther).

When Nikkie joined our colony he was about ten years old. He was then the same size as Dandy, a male of about eight. But whereas Nikkie exploded into growth when he was twelve, Dandy did not, with the result that Nikkie is now almost twice as big. Dandy is Nikkie's exact opposite. He is slightly built, and his eyes have a sensitive, intelligent expression. Dandy is the intellectual of the family. Everyone believes him to be the cleverest in the group. He makes a complete fool not only of the other apes but of humans too. The most entertaining instance I have yet witnessed was with a temporary keeper. The keeper had been grumbling about Dandy for some days because he found it so difficult to coax him outside in the morning. Dandy flatly refused to go out at the same time as the other apes. I advised the keeper to keep Dandy in for a day without any food, by way of punishment. This severe measure had worked successfully on previous occasions. But the keeper thought up what he considered to be a much cleverer plan. A few days later he proudly showed me the result. The other apes were already outside, and Dandy was sitting inside with his hand held up. The keeper placed two bananas in Dandy's hand, whereupon Dandy rapidly went outside. The keeper thought that he had taught Dandy to go out, but to my mind it was much more a case of Dandy's having trained the keeper to give him bananas. I shuddered to think what was in store for us if the other apes were to get wind of the possibilities offered by such blackmail.

Dandy's great intelligence has been evident on a number of occasions. For instance, we have never had to deal with an escape of chimpanzees where Dandy has not been one of the party. This suggests that he is the mastermind behind the escapes, and in most cases we know this to be a fact.

Dandy's social position is such that he has to consider all his actions very carefully. Adolescence in male chimpanzees lasts several years. They are sexually mature at about eight, but they cannot be called socially mature until about their fifteenth year. During these transitional years the male distances himself more and more from the females and children, but he is not yet accepted by the adult males as their equal. In their natural environment these adolescent males often roam around on their own. Sometimes they spend days on end with their mother and younger kin. At other times they hesitantly make overtures to a group of adult males. The adolescent males are fascinated by their elders and betters but receive only harsh treatment and rebuffs at their hands. Until they succeed in winning a place for themselves in the male hierarchy they are in the unenviable position of belonging neither to one camp nor to the other. Dandy, unlike Nikkie, is still in the throes of puberty. He is at a disadvantage compared to his wild contemporaries in that he no longer has a mother to turn to and he is unable to avoid the roughness of the adult males. One of the adult females, Spin, seems to provide him with much-needed maternal warmth and affection. There are periods, even now that Dandy is almost fully adult, when these two are inseparable.

Dandy has to offset his lack of strength by guile. I witnessed an amazing instance of this together with the German cameraman Peter Fera. We had hidden some grapefruit in the chimpanzee enclosure. The fruits had been half buried in sand, with small yellow patches left uncovered. The chimpanzees knew what we were doing, because they had seen us go outside carrying a box full of fruit and they had seen us return with an empty box. The moment they saw that the box was empty they began hooting excitedly. As soon as they were allowed outside they began searching madly but without success. A number of apes passed the place where the grapefruits were hidden without noticing anything—at least, that is what we thought. Dandy too had passed over the hiding place without stopping or slowing down at all and without showing any undue interest. That afternoon, however, when all the apes were lying dozing in the sun, Dandy stood up and made a bee-line for the spot. Without hesitation he dug up the grapefruits and devoured them at his leisure. If Dandy had not kept the location of the place a secret, he would probably have lost the grapefruits to the others.

Nikkie

This experiment was inspired by the methods applied by Emil Menzel, in his study of the transfer of information among chimpanzees. From his studies we knew that the apes were capable of deceiving one another. What we had not expected, however, was the perfection with which the deception was carried out. Dandy's resolute return to the hiding place took us so completely by surprise that Peter Fera was too late to film the incident.

Female Subgroups

Within the group of nine females, one can distinguish three subgroups. A subgroup is formed by females who are often together, who look after one another's children and who support and comfort one another when there is trouble. The largest subgroup is formed by Mama (and Moniek), Gorilla (and Roosje), Franje (and Fons), and Amber. Of this group the last two females have not yet been introduced. Possibly Franje, with her bad teeth and poor health, is already elderly. She is very hesitant and timid by nature. She is always the first to raise the alarm by barking loudly when something disturbs her—when she meets a big spider, for instance, or when, among the hundreds of visitors' faces, she suddenly recognizes the vet. Generally the others pay little attention to Franje's cries of alarm, just as scarce attention is paid to alarm signals given by young apes. (By contrast, an alarm raised by an adult male or high-ranking female causes an instant reaction.)

When Franje is upset, for example after she has been pursued by a male, her legs literally shake, and sometimes she vomits. She evades all trouble and will only interfere in a fight with others if her son, Fons, is involved.

For years Fons was Mama and Gorilla's pet when they did not have children of their own. Perhaps the protection given him by these two influential females explains why Fons shows no sign of his mother's nervousness. Both in appearance and character Fons seems to take after Luit. He has a happy and remarkably friendly nature.

Amber joined this first subgroup when Mama appeared in the colony with her daughter Moniek. Amber seems completely obsessed by this baby. She had to be very patient because Mama did not allow her to handle Moniek until the baby was fifteen months old. Amber was allowed to walk five meters with Moniek sitting on her back before Mama would retrieve her child. As time went on the distance she was allowed to go increased, so that months later Amber had taken upon herself much of Moniek's daily transport and care. Amber became a second mother: an "aunt."

Amber is still young. She is the oldest of the four "girls," that is, the females who came into the group as juveniles. Amber was probably about five years old, while the youngest girl, Henny, was only three when she came here. Over the years these young females have reached puberty one by one. In 1976 Amber had her first genital swelling. With each cycle this became larger and more attractive to the males. Her first pregnancy ended with a miscarriage and the second turned out to be a pseudo-pregnancy. This might seem very disappointing, but in young females such failures are the rule rather than the exception. This is known as a period of adolescent sterility that postpones the big step to motherhood. Now Amber is about eleven, an age when wild chimpanzees may expect their first child.

During adolescence females have an easier time of it than males. They do not have to fight their way into the adult structure, and they are treated with much more leniency than young males. Not only Amber,

Franje and Fons

Luit follows the movements of little Fons, who is barely one year old. Later in life, Fons began to look so uncannily like Luit that there can be little doubt who fathered him.

but the other three girls as well, are fascinated by the children of the other females. The affiliation with the older females works smoothly through the common interest in the youngest chimpanzees. In this way, the technique of child care is transferred from the older to the younger females.

Many people think Amber the sexiest female because of the flirtatious way she moves her hips when walking. But it is doubtful whether this has an erotic effect on the male chimpanzees. She has large, clear, amber-colored eyes and a firm character. More and more she shows the kind of resolution we also recognize in Gorilla. Although Amber's contribution to the group is minor as yet, we can already describe her as a dominant character.

The female subgroup of Mama, Gorilla, Franje, and Amber has roots going back a very long way: the three older females all came from the same zoo. Another subgroup consists of three females who had been together in the Arnhem zoo before the colony was established. One of them has two offspring, and the other two act as "aunts" for these children. The mother, Jimmie, is the least reliable ape where humans are concerned. When unfamiliar people are allowed near the sleeping quarters, Jimmie always tries the same dirty trick to lure them. She pokes a blade of straw through the bars and looks up at the stranger with a perfect poker face. The stranger takes the straw, thinking that this is a friendly gesture, a present. At that moment Jimmie's other hand flashes through the bars and grabs hold of her victim. Then the only way to loosen her grip is with someone else's help.

Toward her fellow apes Jimmie is not so disagreeable. She has an even, almost dull temperament and is on excellent terms with almost all the

Moniek on "Aunt" Amber's back

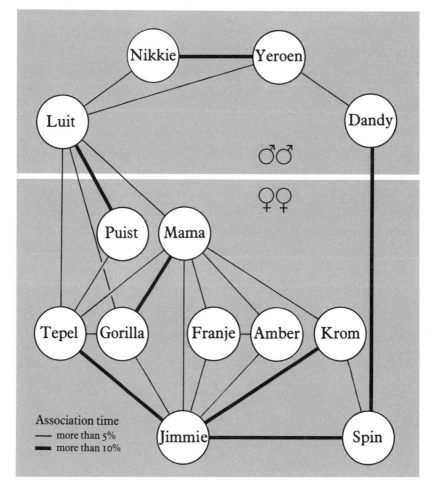

Patterns of Association

By making regular checks we are able to calculate the association preferences, or friendships, among the chimpanzees. This sociogram is based on 2,400 records made between 1976 and 1979. It is limited to the adult group members: at the top the four males, at the bottom the nine females. Narrow lines connect the individuals who spent more than 5 percent of their time within arm's reach of one another. Heavy lines connect individuals whose association time exceeded 10 percent. The maximum was 19.5 percent, between Krom and Jimmie.

The sociogram demonstrates how Mama and Jimmie formed key links in the female relationship network. Puist and three of the four males had little connection with this network. The great exception among the males was Luit, who maintained almost as much contact as the two central females.

others. Her position in social life is at least as central as Mama's. The difference is that Jimmie has much less influence on group processes than the old matriarch.

Jonas, Jimmie's elder son, is the very picture of a spoiled child. While she was pregnant for the second time, she started weaning him. He was then two years old. The weaning made him protest even more loudly than is usual for young chimpanzees. Whenever Jimmie pushed him away from her nipples or held her arm tightly across her breast so that he could not reach, he would burst out in exaggerated screaming, hurling himself down in the sand, rolling and jerking from side to side, and making a noise as if he were choking. Jimmie paid less and less attention to his tantrums, and Jonas had to look elsewhere for sympathy.

For a few weeks Jonas forced Franje to feed him. If she refused and pushed him away or hit him, the monster started to scream, whereupon Jimmie immediately bore down on the unfortunate Franje, threatening and barking. She then stood by Franje until Jonas had been allowed to drink for a while. All Franje could do was stay out of Jonas's way as much as possible.

Almost a year later, after Jonas's brother Jakie was born, it appeared that Jonas still wanted very much to be suckled. Spin had milk to spare after giving birth to a baby that she had, as usual, not accepted. Since Spin has always had a bond with Jimmie and her children, she allowed Jonas to drink. We removed her from the group for a long time hoping she would stop lactating, but it did not help. By the time he was five, Jonas was completely orientated on Spin and spent eight times as long with her as he did with his own mother. Although there was no more milk to be had, he often took his "aunt's" nipple in his mouth. He let himself be cuddled, carried, and protected by Spin. Compared to his peers in the group, Jonas had become a real "mother's baby."

Two more adult members of the group remain to be introduced. One of these is Krom, a female who is inseparable from Jimmie. No other friendship in the group is as closely knit as theirs, not even the tie between Mama and Gorilla, or that of Spin and Dandy.

Krom means "crooked." Her body is distorted, and she has a hunched up way of walking. This can sometimes lead to amusing scenes. The young apes, who think up new games all the time, once had an "ape Krom" craze. For days on end they would walk behind her, single file, all with the same pathetic carriage as Krom.

Four pairs of eyes fixed on an approaching adult male: left to right, *Oor, Amber with Moniek, and Fons.*

Krom has a further handicap: she is deaf. When there is a commotion in the group, she often reacts later than the others. She first has to *see* that something is going on or gather it from the reaction of the other apes in her vicinity. However, despite this handicap, she holds her own excellently in the group. Visual communications, in the form of gestures and facial expressions, obviously supply her with enough information about the relationships within the group. Krom does vocalize herself. The entire, varied repertoire of her species is at her disposal, although her voice sounds a bit strange. Krom's deafness does not stop her functioning normally in a social context, but it is fatal for her offspring. We have let her try repeatedly, but Krom's children have all died within a few weeks. Chimpanzee babies produce all sorts of sounds that are important to the mother. When Krom, for instance, sat on her child, it would start screaming. Normally such protests lead to a hasty correction by the mother, but with Krom no readjustment was forthcoming. Now we take her babies away from her at birth. The first one to whom this happened was Roosje, who is being brought up by Gorilla.

There is a second adopted child in the group: Wouter. He was named after the Swiss primatologist Walter Angst ("Wouter" is the Dutch equivalent of the German "Walter"), who visited Jan van Hooff at his laboratory in Utrecht when I was working there on macaques. Jan invited me to join them on a visit to Arnhem, and it was then that I first saw the chimpanzee colony.

On the day of our visit Spin had given birth. She refused to accept the child, and turned her back on it for hours. The child was removed and wrapped up in towels. It was a boy, and we decided to name him Wouter.

For some weeks Wouter was cared for at Jan van Hooff's home, until news came from Arnhem that another female, Tepel, had given birth. Within an hour, however, her infant had died—possibly because it had been born prematurely. Wouter was quickly taken to Arnhem and put down in the straw of one of the night cages. Then Tepel was let in. To everyone's relief she accepted him promptly. Tepel had plenty of milk to

Jonas still loves to be spoiled and petted by "Aunt" Spin.

Chimpanzees learn what to eat by watching their elders: Jonas (left) closely follows how Gorilla extracts insects from rotten wood.

offer, and Wouter had little difficulty in finding out where it was to be had. Tepel has strikingly large nipples, hence her name, which means "teat." As far as we know, this was the first wholly successful case of adoption by an ape. It was thanks to this favorable experience that I later dared to attempt the more tricky procedure with Gorilla and Roosje. By adopting Wouter, Tepel became the pioneer mother of the colony. She was the first to raise an infant successfully, and she set the example for the other females in the group. Chimpanzees need that. Unlike cats or birds, they do not automatically have an adequate knowledge of parental care. Tepel had brought her knowledge with her from another zoo.

At the age of six, Wouter has the same lanky figure as his natural mother, Spin, whose name means "spider"; they both have unusually long and thin limbs. Wouter also has Spin's undaunted character. She always holds her own indignantly, however mighty her opponents may be. In comparison with Jonas, Wouter is the picture of independence. I have always felt a certain symbolic tie with Wouter, perhaps because I held him in my arms as a newborn baby during my first visit to the colony. He is the most audacious chimpanzee of them all. He throws stones at the observers, the public, and especially at the other apes, and if he provokes an aggressive reaction he flees to his "aunt" Puist.

The third and last subgroup consists of Tepel and her two children, Wouter and Tarzan, who together have a very special tie with Puist. They do not associate unusually often with this large female—perhaps because Puist spends so much of her time with the adult males—but the bond between them becomes particularly apparent during emergencies. Then Tepel, and especially her children, will appeal to Puist for help. As a rule Puist shows no solidarity with females and children, but for this family she makes an exception. The tie is also expressed occasionally when Puist is in a playful mood. Then she skips around with Tepel's children behind her.

In the future another female subgroup may evolve around Amber. The other girls will probably tighten their bond with Amber as soon as she has her first child. These three individuals, Oor, Zwart, and Henny, play such a minor role in the rest of the story that there is no need to introduce them fully—although each already has a personality of her own.

Group Composition

All the chimpanzees' names, except those of animals born into the colony, begin with a different initial. These initials are used as codes in the observations. In this way the composition of the group is also quickly summarized. There are three adult males (Y, L, and N), one adolescent male (D), eight adult females (M, G, F, J, K, S, T, P) and four girls, one of whom is almost adult (A) and three who are adolescent (O, Z, and H).

The names of the apes born into the colony all start with the same letter as their mother's name, except in the case of the two adopted children. Each female's first child has an "o" as the second letter in its name and her second child an "a" (for example, Jimmie's two sons are called Jonas and Jakie). There are seven children in the colony. Only the two youngest are females.

Compared to chimpanzee communities in the wild, this is a small group. Yet groups in the wild are so variable that the size and composition of the Arnhem colony certainly does not fall outside the range. The Japanese primatologist Yukimaru Sugiyama and a colleague followed a community for months that consisted of only twenty-one chimpanzees. The ratio of adult males to adult females did not differ much from that in our colony.

This small feral community was compact, with the senior members spending almost 20 percent of their time together in one group. This is a high percentage for wild chimpanzees who generally disperse in small subgroups, called "parties," throughout the jungle. It is especially on this point that our group is unusual: it spends 100 percent of its time together as one unit.

Shortly before Sugiyama left for Africa to collect this data, he visited Arnhem. He almost failed to leave on his trip at all. While he was leaning out of the window of our observation post, absorbed by the chimpanzees, Nikkie began an intimidation display. Rather late I noticed that Nikkie was holding a heavy piece of wood behind his back. I shouted to Sugiyama in Dutch: *"Kijk uit!"* He nodded, smiling politely, and asked the meaning of this shout. At that moment Nikkie's projectile flew up. With a jerk I pulled Sugiyama away from the window, and the chunk of wood whizzed past his head, just missing him. For the rest of the afternoon he kept a respectful eye on the missile. Later we received a postcard from Africa on which he sent his regards to Nikkie.

As far as we are concerned, Sugiyama achieved a record. Within one day he learned to recognize almost all the apes and was able to point them out and name them correctly. This calls for keen powers of observation, a good memory, and the right attitude: the observer must appreciate that a colony of chimpanzees does not consist of an anonymous mass of black beasts, but that each animal has a unique personality and distinctive appearance. The chimpanzees themselves discriminate between individuals just as we do between people, both within their own circle and outside it. They notice practically every familiar human face, even if it appears among an immense crowd of visitors. Every new face at our observation post, which the apes regard as part of their territory, can count on a small test. Sugiyama's reception was only a little more dramatic than usual.

TWO POWER TAKEOVERS

A HEAVY STEAM ENGINE, AN ADVANCING TANK, AN ATTACKING RHINOC-
eros; all are images of contained power ready to ride roughshod over
everything in its path. So it was with Yeroen during a charging display.
In his heyday he would charge straight at a dozen apes, his hair on end,
and scatter them in all directions. None of the apes dared to remain
seated when Yeroen approached, stamping his feet rhythmically. Long
before he reached them they would be up, the mothers with their children on their backs or under their bellies, ready to make a quick getaway.
Then the air would be filled with the sound of screaming and barking as
the apes fled in panic. Sometimes this would be accompanied by blows.

Then, as suddenly as the din had begun, peace would return. Yeroen
would seat himself, and the other apes would hasten to pay their respects to him. Like a king he accepted this mass homage as his due, obviously regarding some of his subjects as unworthy even of a glance. After
these "formalities" everyone sat down quietly again, the children wandered away from their mothers, and Yeroen relaxed and allowed himself
to be groomed by a group of females or had a brief rough-and-tumble
with Jonas and Wouter, who were always ready for a mock fight with
the big boss. They would chase him and pelt him with sand and sticks,
as if they had lost all sense of deference towards him.

When the existing dominance relationships are suspended in play
there is no danger of confusion arising, because these relationships are
obvious enough at other times. A special form of greeting exists among
chimpanzees which confirms the social hierarchy in a way that leaves
no room for doubt. Before I go on to describe the power takeovers, I will
explain these more or less formal dominance relationships.

Formal and Real Dominance

From the beginning of 1974 until mid-1976 it was quite clear who was
at the top of the hierarchy. At first sight Yeroen's supremacy seemed to
rest on unparalleled physical strength. Yeroen's bulk and his self-assured
manners gave rise to the naïve assumption that a chimpanzee community is governed by the law of the strongest. Yeroen looked much
stronger than the second adult male, Luit. This false impression was cre-

Yeroen (left) *attempts to enlist Nikkie's help by standing beside him and*
screaming at Luit (out of picture).

ated by the fact that in the years of his supremacy Yeroen's hair was constantly slightly on end—even when he wasn't actively displaying—and he walked in an exaggeratedly slow and heavy manner. This habit of making the body look deceptively large and heavy is characteristic of the alpha male, as we saw again later when other individuals filled this role. The fact of being in a position of power makes a male physically impressive, hence the assumption that he occupies the position that fits his appearance.

The impression of a connection between physical size and social rank is further strengthened by a special form of behavior that is the most reliable indicator of the social order, both in the natural habitat and in Arnhem: the *submissive greeting*. Strictly speaking, a "greeting" is no more than a sequence of short, panting grunts known as pant-grunting. While he utters such sounds the subordinate assumes a position whereby he looks up at the individual he is greeting. In most cases he makes a series of deep bows that are repeated so quickly one after the other that this action is known as bobbing. Sometimes "greeters" bring objects with them (a leaf, a stick), stretch out a hand to their superior, or kiss his feet, neck, or chest. The dominant chimpanzee reacts to this "greeting" by stretching himself up to a greater height and making his hair stand on end. The result is a marked contrast between the two apes, even if they are in reality the same size. The one almost grovels in the dust, the other regally receives the "greeting." Among adult males this giant/dwarf relationship can be accentuated still further by histrionics such as the dominant ape stepping or leaping over the "greeter" (the so-called bluff-over). At the same time the submissive ape ducks and puts his arms up to protect his head. This kind of stuntwork is less common in relation to female "greeters." The female usually presents her backside to the dominant ape to be inspected and sniffed.

The presentation and inspection of the female genitals is a characteristic form of chimpanzee behavior but one that, to judge by the number of comments made by members of the public, arouses a strong reaction in humans. I once read that presentation of one's backside might prove useful if one were being attacked by an adult male chimpanzee: "Pull down your knickers and show him your bare bottom!" Although I would not advise anyone to try such a tactic, it is quite possible that presentation by female chimpanzees does have an appeasing function. But they never do it at the actual moment of attack. Once the aggressor charges at

Nikkie bows to Yeroen and "greets" him with a series of panting grunts. Yeroen ignores this mark of respect. (Luit is on the left.)

them it is too late for submissiveness. The only choice that remains is between fleeing and fighting. In order to avoid such a situation it is necessary for low-ranking apes to detect an aggressive mood before it develops too far. If the potential aggressor is a female, this margin of safety will usually be completely absent: a female's fury can burst forth without any warning at all. With the males, however, several minutes will elapse between the first slow swaying movements of the upper part of their bodies, the raising of their hackles, the gradual increase in the volume of their hooting, and the culmination in attack. An attempt at pacification by means of "greeting," grooming, or presentation will have to be made before the male reaches the point of no return and starts his attack.

Submissive "greeting" is seen when apes meet each other, as a reaction to the approach of a dominant group member or the first signs of display, and above all a short while after a charging display or aggressive explosion. "Greeting" is a kind of ritualized confirmation of the dominance relationships. Such confirmation may also be seen after

Quarrels between females do not have the slow escalation to a climax that is characteristic of male quarrels. Female aggression is more impulsive. Amber (left) screaming after an unexpected slap by Gorilla.

conflicts in which the tables have been turned temporarily. The dominant ape sometimes comes off worse. It is quite normal among chimpanzees for low-ranking group members to protest aggressively at being slapped, and they regularly put the dominant ape to flight or even physically overpower him, especially when they combine their forces. Yeroen hardly ever found himself in such a predicament, but I have seen him being chased on several occasions by a group of furious, screaming females. It was a revelation to me that even the self-assured Yeroen could be frightened. Although he was not so frightened that he bared his teeth or screamed during his flight, the image of the almighty, invincible leader was none the less undermined. But such events did not affect his position permanently because during the ensuing reconciliation he was "greeted" as usual by the females.

Hence dominance manifests itself in two very different ways. First, there is social influence, or *power,* as reflected in who can defeat whom and who weighs in most heavily when a conflict in the group occurs. The outcome of these confrontations is not 100 percent predictable, particularly since chimpanzees constantly form shifting alliances. Incidental reversals in the social hierarchy are far less rare than with other animal species. That is why the chimpanzee hierarchy is so often termed "flexible" and "plastic." A young chimpanzee of not more than two or three can sometimes put an adult male or female to flight or even coerce them into doing something. These are not just playful incidents; they can be serious conflicts, such as the occasion when Jonas, with his mother's support, forced Franje to suckle him.

Children are never "greeted" by adult group members: they may sometimes exert real power but they have no *formal dominance.* Unlike the outcome of conflicts, which are so variable that even the leader is sometimes chased up a tree, the "greeting" rituals are completely predictable. "Greeting" reflects *frozen* dominance relationships. It is the only common form of social behavior that is non-mutual: in other words, if A "greets" B during a certain period, B will *never* "greet" A during the same period. This remarkable rigidity exists only for the submissive greeting, in which a series of low, panting grunts is emitted. Chimpanzees greet each other in many different ways. I shall use the term "greeting" in quotation marks to denote the vocal form. Yeroen, as the alpha male, never uttered pant-grunts, but he was frequently "greeted" by everyone in the group.

Formal rank and power generally overlap; however, in some cases, rank can become dissociated from power. In other words, the position of a dominant ape can become untenable. Just how chimpanzees determine this moment is not known, but it seems clear that the course of their aggressive encounters constitutes the main source of information in this respect. For example, if the subordinate party begins to win conflicts more and more often, or if he at least regularly produces fear and hesitation in the dominant party, this will not escape him. If this shift in the relationship persists, the "greetings" between them will gradually become no more than hollow formality. The subordinate party will stop "greeting" the dominant party. In this way he queries, as it were, the state of their relationship. This first step—ceasing to "greet"—was observed in all the reversals of dominance in the Arnhem colony. In the spring of 1976 Luit summoned up the courage to confront Yeroen with this challenge. Their relationship exploded, and the whole group was plunged into a period of reorganization that took a year to complete.

Yeroen was at one time so all-powerful that he alone received more than three-quarters of the "greetings" in the group; during some periods this figure even rose to over 90 percent. Luit also "greeted" him often and was himself "greeted" far less by the others. High-ranking females, such as Mama and Puist, never "greeted" Luit. If we add to this that Luit looked a lot smaller and weaker than Yeroen and always kept very much in the background, it will be obvious that a coup by Luit was the last thing I expected.

The First Blow

The summer of 1976 was extremely hot and dry. All over Europe the grass gradually turned brown. The woods around Arnhem were ravaged by huge fires. At one point a fire came so close to the zoo that we feared for the safety of the animals. Within the chimpanzee community it was truly a long, hot summer in the social sense as well. On the afternoon of June 21, I saw Yeroen bare his teeth for the very first time. It was also the first time I heard him scream and yelp, and the first time I witnessed him in need of support and reassurance. Moreover, the difference in size between him and Luit had suddenly vanished.

It had been obvious in the morning that significant changes were afoot, because Luit had openly mated with Spin. Spin had just come into her estrus period—her genital swelling was prominent and she was sexually attractive to males. Yeroen was normally extremely intolerant of other males having sexual intercourse, but on this occasion he lay stupidly in the middle of the field and made no move when Luit and Spin mated just 10 meters away from him. He even went so far as to turn his back on them, as if he preferred not to watch this distasteful scene. Our initial assumption was that Yeroen was seriously ill and that Luit was taking advantage of the situation. We were proved wrong, however, that evening, when Yeroen's appetite was found to be perfectly normal.

In the afternoon Luit had displayed in large circles around the old leader and had provoked several fascinating interactions. This way, he gave the starting signal for a protracted series of impressive displays and conflicts between himself and Yeroen that increased in intensity day by day. Here follows a report of the first open confrontations:

21 June 1976: 1:45 P.M.

Luit circles Yeroen with all his hair on end at a distance of about 15 meters. He stamps his feet and thumps the ground with the palm of his hand. He picks up any sticks or stones he finds in his path and hurls them away. Yeroen is sitting on the grass and glances swiftly up at Luit every now and again. When the challenger is behind him Yeroen does not turn round, but moves his head slightly so that he can unobtrusively watch what Luit is doing through the hair on his shoulders. Sometimes Luit comes within a few meters of Yeroen. Yeroen then stands up, hair on end, and takes a step forward. During this brief confrontation the two males do *not* look at each other, and, as soon as Luit has passed by, Yeroen immediately returns to his spot on the grass.

After having completed six or seven of these circles Luit walks over to Spin and inspects her sexual swelling. Gorilla walks over to Yeroen and gives him a kiss. Both Nikkie and Dandy walk over to Spin, passing Yeroen on the way. Neither of the apes "greets" Yeroen, which is highly unusual. Luit is playing with Fons, at that time the youngest child in the group. Fons's mother, Franje, and Mama move over to sit by Luit. Luit starts grooming Mama. For a long time after this the group is quiet. The situation is unchanged until Yeroen gets up and starts to go over to Spin.

2:25 P.M.

As is her habit Puist is in the company of the sexually attractive female, in this case Spin. Puist "greets" Yeroen, who goes over to join her and Spin. Luit immediately becomes restless. His hair is slightly on end;

Luit displaying in large circles around Yeroen.

he starts gathering some large sticks and begins to hoot softly. At this Yeroen leaves Spin and walks past Luit displaying slightly, once again without looking at him. Yeroen goes up to Mama and starts to groom her. While Yeroen is doing this Mama keeps looking around to see what Luit is doing. Luit embraces Franje for a short while before going and sitting down right in front of Yeroen and Mama while he builds himself up for more overt actions. Luit's hoots gradually become louder until eventually he stands up and rushes full tilt at Mama and Yeroen, narrowly missing them. Yeroen had momentarily stood up but quickly sat down behind Mama's back as Luit rushed past.

2:35 P.M.

While Luit is preparing his second action against Yeroen, Nikkie takes advantage of the situation to mate with Spin. No one seems to notice what has happened. Luit first charges at Mama who flees screaming. Then Luit sits down in front of Yeroen and hoots defiantly at him.

Yeroen is sitting alone under a sloping tree trunk, which Luit proceeds to climb, displaying and hooting loudly as he does so. Yeroen looks up at him hesitantly. Luit is now a few meters above his rival and thumps rhythmically and forcefully on the tree trunk. Finally he leaps down, landing right next to Yeroen. Luit gives Yeroen a resounding smack and immediately runs away. Yeroen bursts out screaming at the top of his voice. He runs to a group consisting of Gorilla, Krom, Spin, Dandy, Henny, and some other apes and embraces them all in turn. Pandemonium breaks loose involving nearly all the apes. Backed up by a large group of hooting, screaming, and barking supporters-cum-sympathizers Yeroen approaches his challenger.

Until that moment Luit had been watching him from a distance with his hair slightly on end, but now he flees screaming in the face of Yeroen and his band. Aggressive sounds are heard from all directions, and Luit finds himself attacked by a crowd of ten or more apes. Some members of the group, however, refuse to become involved. Jimmie, for example, keeps her distance. Twice Yeroen goes up to her and holds out his hand, yelping, but on both occasions Jimmie turns and walks away from him. The crowd continues to pursue Luit for a few minutes and then stops abruptly. Silence falls, broken only by Luit's screams. He has been driven into a corner on the far side of the island. Luit has clearly lost the first battle.

Oor, one of the "girls," goes up to Luit and presents herself to him. Luit presents himself in return, still screaming, so that both of them stand briefly bottom to bottom. Suddenly Luit has a tantrum. Just like a chimpanzee infant who fails to get his way, he rolls over and over in the sand screaming, beating his head with his hands, and emitting choking sounds as if he were going to be sick. Oor goes over to him once again and embraces him. Bit by bit Luit calms down and then proceeds slowly to follow Yeroen as he walks back to the middle of the enclosure.

The whole incident has lasted no more than five minutes. Apart from the initial smack there was no physical aggression.

2:40 P.M.

A few minutes later Luit is already displaying slightly once again, and he launches an attack on Puist, who is sitting with Spin. Having done this he invites Spin to mate. The moment Spin accepts his invitation Yeroen begins to scream, but he does not dare to intervene. Dandy reacts by "greeting" Yeroen and embraces him in an effort to calm him down. All the while Luit mates undisturbed with Spin.

2:50 P.M.

Yeroen has joined Mama and Franje and is tickling little Fons playfully. He has a playface but is nevertheless not completely at ease, because he keeps shooting glances in Luit's direction. Luit is sitting swaying to and fro a short distance away, picking up twigs with his hands and breaking them with his teeth. Mama's reaction to Luit's behavior is to leave Yeroen. Yeroen follows her, but when he tries to sit down beside her Mama gets up and moves away again. This pattern is repeated several times. Yeroen tries to sit down next to Mama, but Mama makes it quite clear that she does not want his company. Meanwhile Franje, the other female who had been sitting with Yeroen, is attacked by Luit. He starts jumping up and down on her back. Both Franje and her son, Fons, begin to scream. Fons is clinging to Franje's belly and is being squashed between his mother and the ground each time Luit jumps on them both. Luit then switches his attention to Yeroen and Mama and displays in their direction several times. Mama keeps walking away from Yeroen, but Yeroen follows her, until Luit finally succeeds in chasing Mama right away.

All the other apes have also dispersed, and peace descends again. The two male rivals are now left facing each other, all alone in the middle of the enclosure, about 6 meters apart and deliberately avoiding each other's gaze. Luit watches some pigeons fly over, and Yeroen stares at the ground.

3:10 P.M.

The final stage of this unusual series of events is triggered by Krom, who walks over to Luit, "greets" him, and begins to groom him. Luit, however, gets up and walks away, and Yeroen does the same. Krom, imperturbable as always, goes up to Luit and starts grooming him again. At first Yeroen moves about 10 meters away, but then he hesitantly turns back and slowly approaches Krom and Luit. Krom then goes to "greet" Yeroen who grins nervously and embraces her. Luit goes to the spot where Yeroen was first sitting, sniffs the ground, and sits down himself. Strangely enough Yeroen answers this by going and sitting down in Luit's old spot, and so the two males sit facing each other again.

Now Krom begins to groom Yeroen, and Luit starts displaying round them. He makes a series of wide circles and only stops after both he and Yeroen have hooted loudly in unison. Luit moves closer, and Krom immediately leaves Yeroen and goes over to groom Luit. This is the fourth time she has groomed one of the two males. This situation goes on for a good five minutes: Yeroen sits alone while Krom and Luit sit not far away grooming each other.

Suddenly, Luit leaves Krom and hesitantly walks up to Yeroen. Both males have their hair on end, and for the first time they look each other straight in the eye. Yeroen briefly embraces Luit, whereupon Luit presents to him and allows Yeroen to groom his backside. In the case of reconciliations between adult males it is quite normal for the bottom to be groomed first before they move on to other parts of the body. Krom withdraws. It is 3:30 P.M. The two rivals groom each other for about fifteen minutes.

The Isolation of Yeroen

The detailed report above is an account of the first two hours of a process that lasted two months. In the first encounter no decision was reached. We may wonder why the two rivals did not put an end to their conflict by fighting it out once and for all. The answer is simple: because physical strength is only one factor and almost certainly not the critical one in determining dominance relationships.

In my experience about four out of ten bluff displays between adult males end in a conflict such as occurred when Yeroen began to scream and Luit smacked him hard. Typically such incidents involve threats, chases, and screaming. Hitting is rare between males, but a single blow does not in itself constitute a fight. In serious encounters the opponents actually get hold of each other and start biting. About one conflict in a hundred leads to a real fight, that is, 0.4 percent of all confrontations between males. The threat of a fight is, however, always there, and it is this that makes the dominance process so tense.

The social maneuvers in which Luit and Yeroen involved the females were in fact far more sensational than a simple physical confrontation. The process of action and reaction took place relatively quietly. It continued over a longish period of time, but the effect was quite dramatic. Both males repeatedly sought contact with the adult females. Yeroen in particular constantly sought their company, which is not surprising in view of the protection he got from them. Many times throughout the whole transitional period—from the first blow to the establishment of the new leadership—all the females supported Yeroen against Luit, whereas the reverse did not happen once.

It is extraordinary that the nine females appeared to be so much in agreement, although whether they were in fact unanimous remains open to doubt. The struggle for dominance between Yeroen and Luit sometimes created tense relationships between the females. Mama and Gorilla were obviously willing to support Yeroen, whereas other high-ranking females, such as Puist and Jimmie, were less inclined to do so. Puist was even seen to attack Mama on several occasions when the latter took Yeroen's side against Luit. Later, the reverse was also seen when the united female front began to crumble. Once Luit had finally established himself as leader, Puist was the first female to desert Yeroen and ally with the new dominant male. Initially, Mama was enraged at Puist's desertion and would attack her whenever she openly sided with Luit. It is quite conceivable that females such as Puist and Jimmie would have gone over to Luit far sooner had it not been for Mama. The months of concerted female support for Yeroen may have been due more to Mama's overriding influence than to spontaneous unanimity.

One would have thought that with such a powerful group behind him Yeroen had nothing to fear. But it was clear from the first day that he was in danger of losing the group's support. Mama avoided his company on several occasions. This was undoubtedly due to Luit's undermining tactics. Luit became restless whenever he saw Yeroen with an adult female. He would go over to Yeroen and the female and punish the female in no uncertain terms. Even if her contact with Yeroen had only been brief and was already over, she still ran the risk of being castigated by Luit (as Franje was in the episode described above). Once he had begun to use this tactic, on the first afternoon of the struggle, Luit maintained it for weeks on end with steely persistence.

Sometimes Luit only had to stand up and start moving toward the group and contact between Yeroen and the female would cease immediately. On other occasions he met resistance, and the confrontation escalated into a conflict. If Luit had been completely on his own, these conflicts would have been decided without exception to his detriment because his opponents had strength in numbers. The females would

Luit punishing Tepel for having sat by his rival. Yeroen (left) is standing looking on but does not dare defend her.

have had little reason to avoid Yeroen's company out of fear of Luit's aggressive reaction. But the situation was not as simple as that. The interventions of the third large male, Nikkie, meant that it was never certain at the outset whether Yeroen would end up the dominant party and whether it would be risky for the females to become involved in a conflict between Yeroen and his challenger.

I call the actions that Luit displayed against Yeroen's contacts in the group *separating interventions*. Their short-term effect was obvious enough, but, in order to determine whether they also had a long-term effect, we made a statistical analysis at the end of the year. This was necessary because subjective impressions are not completely reliable, particularly when the processes are as slow as these were. Every five minutes we had recorded on a portable tape recorder exactly which individuals in the enclosure had joined together to form a subgroup, that is, were

A separating intervention: Yeroen (foreground right) and Krom (left) were sitting grooming each other but were forced to separate by Luit, who is displaying on a tree trunk above them. Tepel and Jonas look on.

sitting within 2 meters of one another. By analyzing literally hundreds of these five-minute checks made in the summer of 1976, we reconstructed a picture of Yeroen's relationships.

In the spring of 1976, when Luit was still regularly "greeting" him, Yeroen spent about 30 percent of his time in subgroups with adult females. In the weeks preceding Luit's first overt challenge we found that this percentage more than doubled. This meant that in that period Yeroen almost continually sought the company of the females. He probably withdrew into their midst because he already sensed that Luit's attitude was changing and he knew that his position was threatened: in that period Luit only very rarely "greeted" Yeroen. The fact that Yeroen sought the safe haven of female company in the period of calm before the storm is something that was discovered in retrospect, when we analyzed our records. This confirmed that the preparatory moves were made at a time when we ourselves were not yet aware that anything untoward was afoot.

Data from later periods demonstrates a marked change. During the weeks when Luit was actively challenging Yeroen's leadership and making innumerable separating interventions, the amount of time Yeroen spent in the females' company very gradually decreased. By the autumn Yeroen's increase in contact with the females had dropped off, and he was spending even less time with them than he had been doing in the spring. Our data confirmed Yeroen's social isolation.

Luit's negative attitude toward the females was only apparent when they associated with his rival. On other occasions his behavior toward them was extremely positive. He often sat with them, groomed them, and played with their children. These contacts sometimes took place at moments of tactical importance. This was obvious on the very first day, after Luit had been defeated by an alliance between Yeroen and the females and had made his peace with Yeroen. That same afternoon Luit started to display again, and once more his challenge resulted in a major conflict. In the minutes preceding his challenge Luit had "done the rounds." First of all he went to Franje, groomed her, and played with her son Fons for a short time. Then he went to Spin and groomed her, then Gorilla, and then Puist. Immediately after this unusually rapid succession of contacts Luit launched into his intimidation display around Yeroen. Could it be that Luit's behavior toward the four females was an attempt to get them to remain neutral during the subsequent clash? Was it a kind of "bribe" or an attempt to "arouse sympathy" by his friendly actions?

The problems between Luit and Yeroen would never have reached such proportions if it had not been for the females. As individuals the question of their dominance relationship was settled within a week. This was apparent from their behavior in their sleeping quarters, where they spend the nights apart from the rest of the colony. Whereas previously Luit had tended to remain in the background, he now moved around with noticeable freedom. Sometimes he even took away apples meant for Yeroen. On the whole the two rivals were fairly peaceful when they were in their night quarters together, but twice they clashed. On both occasions we found that Yeroen had far more wounds. Although these were not at all serious, Yeroen looked a pitiful sight. He had lost all his former self-confidence, and the look in his eyes reflected the psychological beating he had taken. The first morning Yeroen appeared in such a battered state, the group reacted excitedly. When Mama discovered Yeroen's wounds she began to hoot and looked around in every direction. At this Yeroen broke down, screaming and yelping, whereupon all the other apes came over to see what was the matter. While the apes were crowding around him and hooting, the "culprit," Luit, also began to scream. He ran nervously from one female to the next, embraced them, and presented himself to them. He then spent a large part of the day tending Yeroen's wounds. Yeroen had a gash in his foot and two wounds in his side, caused by Luit's powerful canines. It was small wonder that these wounds caused such a stir, because it was the first time in years that Yeroen had been hurt.

Yeroen pursues his challenger, Luit. Both males are screaming.

Luit was simply stronger than Yeroen. Because of Luit's physical superiority, Yeroen's social isolation in the autumn of 1976 was all-important. Once he was isolated, not only in his night cage but also in the group, Yeroen was bound to lose.

The Open Luit–Nikkie Coalition

Nikkie's role was crucial in the conflict between Luit and Yeroen. Despite the fact that Luit and Nikkie acted independently of each other, they still formed a coalition: an open coalition to be exact. They were not in each other's company all that much, they did not exchange any special signals during simultaneous actions, and the support they gave each other was largely indirect. And ironically enough the fruits of their collaboration constituted a time bomb that threatened ultimately to destroy their relationship. The outcome for Luit was the leadership; for Nikkie it was the position of second-in-command in the hierarchy, above the adult females and above Yeroen. The promotion of Nikkie was in fact even more of a change than that of Luit, because at the beginning of 1976 Nikkie had scarcely been taken seriously at all. None of the adult females "greeted" him, and he was regularly sat upon. No one was surprised or disturbed when Mama beat Nikkie for having been rough with her.

In the period preceding the challenge to Yeroen's leadership Nikkie adopted the habit of acting toward the females in exactly the same way as Luit did. If Luit attacked a female, Nikkie would immediately give the same female a hard slap. This hit-and-run tactic became so characteristic of Nikkie that we regarded him as a cowardly profiteer. But subsequent events in that same year cast a whole new light on Nikkie's practice of playing Luit's shadow. Perhaps the two of them were already testing Yeroen's position because their attacks on the females often took place dangerously close to where Yeroen was sitting. Even females who fled to Yeroen for protection and embraced him were not always safe from Nikkie and Luit. The fact that Yeroen did not act forcefully in such circumstances may have been a significant indicator to the other two males. A leader who hesitated to defend his protégées might very well have problems defending himself.

During the period of open conflict between Luit and Yeroen, Nikkie only once interfered directly in a confrontation between the two rivals,

By the summer of 1976 Nikkie had become a fully adult male.

and strangely enough his action on that occasion was directed against
Luit. It happened early on in the struggle, about ten minutes after Luit
had roughly interrupted a mating session between Nikkie and Spin by
attacking Nikkie. This was an extremely significant incident because
it showed that Luit could not afford to be in direct competition with
Nikkie. On the very first day a similar, but less obvious incident had occurred; once again sexual rivalry was the cause. After having been punished by Luit, Nikkie ran screaming to Yeroen and threatened to form a
coalition against Luit. Luit reacted by hurrying to the other side of the
enclosure. These two incidents may explain why Luit went out of his
way to avoid getting into conflict with Nikkie. So as not to get Nikkie to
turn against him, he had to act tolerantly. Luit could not risk alienating
Nikkie because he needed his help too badly.

Nikkie openly turned against Luit once, but he indirectly sided with him many times by helping to fight off Yeroen's supporters, the females. Without Nikkie's help Luit could not possibly have succeeded in dethroning Yeroen. The normal pattern of interaction would be as follows. Luit would begin by displaying around Yeroen until Yeroen could no longer ignore him and went screaming in search of help. Yeroen would either beg the females to come to his support or he would go and fetch them. As Yeroen and his supporters approached Luit, Nikkie would spring into action and attack one of the female supporters, preferably either Mama or Gorilla. The effect of his intervention would be to confuse the situation; while the conflict between Yeroen and Luit continued, the females would band together against Nikkie. On most occasions Yeroen and Luit would end up high in the dead oak trees, with Luit displaying and Yeroen screaming and holding out his hand in vain to his female supporters down below, who had their hands full trying to deal with the indefatigable Nikkie.

Nikkie was so fast and acrobatic that they hardly ever succeeded in catching him. He would rush about and leap over his opponents, dodging them at every turn, until they no longer knew where to look: sometimes they would receive a slap from behind when they had been expecting a frontal attack. During one such scene Nikkie came up behind Mama, placed both his hands under her large bottom, and threw her up into the air before she knew anything about it. These stunts demonstrated Nikkie's remarkable physical strength; he could no longer be ignored or kept in his place. His fight with the females took on a life of its own, quite separate from the struggle high up in the trees. It sometimes happened that Nikkie and the females were still at odds long after the conflict between Yeroen and Luit had ended. Sometimes, however, Nikkie seemed deliberately to come to Luit's aid. On one such occasion he chased Mama away from Yeroen's side, drove her into one of the small dead trees, and stayed at the bottom of the tree displaying until Luit and Yeroen had finished quarrelling. Then, and only then, was Mama allowed to descend.

Conversely, Luit supported Nikkie in his gigantic struggle against the females. On the rare occasions that Nikkie found himself in trouble either Luit or Puist would rally to his rescue. Here again Puist's role was strangely unfeminine. While all the other females were displaying their solidarity in attacking Nikkie, Puist would be the odd one out, siding with the "wrong" side. Her special relationship with Nikkie, like Luit's

A coalition between Mama (left) and Krom against Nikkie (right).

relationship with Nikkie, was a mutual one. Nikkie's aggressive actions were never directed against Puist and she was the only female who could count on his support if she herself got into deep water.

If Nikkie had thrown himself wholeheartedly into his struggle with the females—that is, if he had used his dangerous teeth—he would definitely have forced them into submission earlier. As it was, he abided strictly by the rules, fighting only with his hands and feet and never biting with his canines. Males sometimes do bite females, but only with their incisors. The females, who do not have large canines, are far less wary of using their teeth, both in fights with other females and when defending themselves against male aggressors. Because of these rules the weeks of conflict between Nikkie and the females showed no definite outcome either way, nor was there any serious fighting, because Nikkie was agile enough to avoid being caught. Time and again he provoked the females, and when they screamed in protest he waited to see how near they dared to get to him. If they got too close they knew they would receive a slap. If, on the other hand, the females outnumbered him, then Nikkie would leap agilely out of the way. Despite Nikkie's strength and the females' indignation, their confrontations never led to any injuries.

The fact that such spectacular dominance processes can take place in a large colony without bloodshed justifies the optimism with which the Arnhem project was started. And at the same time it undermines the argument of zoo and laboratory managements that chimpanzees must be kept isolated, or at most in very small groups, because of their so-called "extreme aggressiveness." Male chimpanzees are incredibly strong and they have the power to kill, but they are also capable of containing themselves. Nikkie had to all intents and purposes a knife in his pocket, but nevertheless fought the females with his bare hands. The use of the canines remains restricted to the rare times when males fight among themselves, and even then chimpanzees are usually governed by a strict code of behavior.

Over several months Nikkie's confrontations with the females gradually decreased as more and more females began to "greet" him. The last females to acknowledge his new position were Mama and Puist. They did not "greet" Nikkie until October. From that time on Nikkie ranked above all the females. He achieved his new status not merely because of his rapidly increasing physical strength (Nikkie grew explosively in 1975 and 1976) but also thanks to the chaotic situation within the group. The females had no one to turn to. The group had no leader, and moreover Nikkie invited conflicts at times when the two candidates for the leadership were busy confronting each other high up in the trees. The great advantage that Luit gained from Nikkie's actions was perhaps only a by-product as far as Nikkie himself was concerned. Nikkie and Luit did not have identical but parallel interests: Luit's bid for power was probably made easier by the struggle for dominance between Nikkie and the females, and vice versa. Because their collaboration was obvious and yet their fields of action were so different, I have called their coalition an *open coalition.*

The only instances of more direct collaboration between Luit and Nikkie were after Yeroen had been virtually isolated within the group. By this time when there was a conflict between Yeroen and Luit, Nikkie accompanied them, in the role of court jester. He did not actually participate in the conflict but watched from the sidelines, running to and fro, turning somersaults and generally encouraging Luit from a distance. He hooted when Luit hooted at Yeroen, and sometimes he would chuck objects at Yeroen when he nervously tried to avoid Luit's advances, yelping as he did so. It was as if Nikkie were amusing himself at Yeroen's expense.

Tantrums and Fights

It may seem as if during the summer of 1976 the chimpanzees did nothing but fight. This impression is wrong. The incidents described were highlights in a torpid season. For hours on end the chimpanzees would sit or loll about in the sun, barely able to keep their eyes open. Occasionally a few apes might lumber slowly around the enclosure or groom each other quietly in the shade. But despite their apparent lassitude the chimpanzees never forgot what was afoot. Whenever Luit sleepily got up to take stock and happened to notice that Yeroen had sought the company of a female while he had been asleep, we immediately focused the video camera on him, ready for action. Three or four times a day all hell broke loose. It is these moments I have recorded, not the remaining hours of deep peace and ostensible harmony.

The period of struggle for dominance between Luit and Yeroen was a tense but exciting one for us. It was a process full of twists and turns. For the first month it was not clear what the final outcome would be. Some days Yeroen seemed to be in control, bluffing Luit aside or chasing him away, assisted by the females. On other days it was Luit who had the upper hand. He displayed forcefully at Yeroen and never bared his teeth once during confrontations with his rival. Yeroen, on the other hand, did bare his teeth, a sign of uncertainty among chimpanzees. In the early stages it seemed as if Yeroen were trying to hide this uncertainty from Luit. Without any definite expression on his face Yeroen would move away from his challenger and would only grin or utter a soft yelp when he was far enough away and had his back turned to Luit. This fascinating spectacle of "keeping up appearances" was even more obvious in later dominance processes.

Another phenomenon, which was also seen in later processes and which I have come to interpret as the beginning of the end, is the losing party's temper tantrums. Yeroen began to have tantrums after the conflict had been raging for about a month. With an unerring sense of drama he would let himself drop out of a tree like a rotten apple and roll around on the ground screaming and kicking, while all the time Luit was displaying. These hysterical outbursts gave an impression of scarcely suppressed despair and abjectness. When he had regained some of his self-composure, Yeroen would run yelping to the females, throw himself down on the ground a few meters away, and stretch out both hands to them. This was not a begging gesture but a beseeching gesture, beseeching them for their support. If the females refused to help, or even went out of their way to avoid him, Yeroen would once again break down and have a tantrum. He seemed to lose all control of his muscles, screamed pitifully, and writhed around like a fish on dry land.

His reaction was very different if the females did offer their support. Then he would leap up immediately, embrace them, and turn on his rival with the females close behind him. But as time went on the females became less and less willing to help Yeroen, which was hardly surprising in view of Nikkie's hard-handed interventions. As his sense of powerlessness increased in the face of Luit's challenges, so did Yeroen's tantrums. It was as if Yeroen were trying to arouse pity and mobilize his sympathizers against Luit. Familiarity, however, breeds contempt; in time Yeroen's tantrums became commonplace and predictable, and the other apes ceased to take any notice. The same was true of us. To begin with we stood rooted to the spot, full of pity for Yeroen's "inconsolable despair." But in time we became more laconic. We found it difficult to take Yeroen's despair completely seriously: it seemed so exaggerated and

Wouter embraces Yeroen in an attempt to comfort him during one of Yeroen's many temper tantrums.

posed. Yeroen stopped having frequent temper tantrums once he had lost virtually all support. After a confrontation with Luit he no longer uttered heart-rending screams but sat staring at nothing in particular. The stuffing had been knocked out of him.

The interesting thing about the temper tantrums was that Yeroen, at the ripe old age of thirty, attempted to attract attention and arouse sympathy by lapsing into childlike, or rather childish, behavior. Tantrums are normally associated with infants when they are being weaned. Children feel that their mother is rejecting them and scream and kick until she draws them to her again. It is surprising (and suspicious) how abruptly children snap out of their tantrums if their mother gives in. It may well be that Yeroen displayed the same kind of behavior because he felt frustrated and threatened by Luit's campaign to dethrone him. Yeroen was, as it were, being *weaned* from power.

Yeroen's slow fall from dominance was reflected in his most serious clashes with Luit. I have already mentioned their two fights in their sleeping quarters, both of which were won hands down by Luit. If we call the day the struggle for dominance broke out day 0, then the night fights occurred on the 30th and 59th nights. Other serious incidents were also observed during the day. These are described below. For the purpose of orientation I regard the dominance process as having come to an end on the 72nd day, when Yeroen "greeted" Luit for the first time.

Day 16

In the first physical fight between the two males Yeroen is the attacker. In reaction to Luit's provocative displays Yeroen, helped by Gorilla, chases after Luit and catches him. Luit tries to break loose but finds himself surrounded by enemies (at least eight individuals are involved in the fight). Before Luit finally flees up a tree Yeroen succeeds in biting him one last time. Luit does not bite back at all.

Day 24

Once again Yeroen is the aggressor, provoked by Luit's displays. Together with Gorilla and Puist, Yeroen chases Luit at full speed, screaming as he runs. Luit is also screaming. Just when Yeroen succeeds in catching up with Luit, Yeroen accidentally (?) goes head over heels. After this the chase continues but ends abruptly when both males flee up a tree and the two females do not follow. This is not altogether surprising because the tree the males have chosen is a live one, protected by high voltage electricity. The conflict has become so intense that the males have dared to brave the electric shocks. While the two rivals sit up in the tree the whole group gathers down below to gaze at this amazing spectacle. Yeroen is the first to jump down and is comforted by several of the group. They kiss and embrace him, and some of them put their fingers in his mouth. Yeroen repeatedly invites Luit to come down. His invitations are accompanied by a whole scale of gestures: from holding out his hand and pouting his lips, right up to sexual advances (Luit and Nikkie also have erections). But Luit remains in the tree and breaks off a number of branches, which he throws down. The other members of the group pick them up and start to eat the leaves. Several times Luit hangs just above Yeroen's head, balancing on a few thin branches, and holds out his hand to him. It is three-quarters of an hour before Luit finally descends with his arms full of broken branches. While the other members of the group gather greedily around the food Luit has brought down with him, Yeroen follows Luit to another part of the enclosure where they are quietly reconciled. Once they are friends again the two groom each other for a considerable time and then settle down in brotherly fashion to feast on the sizeable branch Yeroen has been holding all along.

This dramatic event had far-reaching consequences because from that day on the apes' interest in the live beech trees greatly increased. The next day Luit was seen sitting under the same tree on several occasions, looking up at it, although he did not venture to climb up again. Four days later the apes had solved the problem of the electric wires by placing a long branch against the tree as a sort of ladder. This method was later used more frequently and developed into a speciality of Luit's and Nikkie's. Yeroen never showed any initiative in that direction. It may have been accidental that Luit played Santa Claus on that first and later occasions, strewing food around for the masses, but to me it immediately looked as if he had hit on a very clever way of drawing attention to himself.

Day 36

Another attack by Yeroen after repeated provocation by Luit. Yeroen corners Luit in the top of one of the (dead) trees and bites him. This time Luit bites back. Mama, Gorilla, Puist, and Nikkie also play a role in the

Yeroen holds out a hand to Luit, inviting him to come out of a live tree after a conflict between them. Left to right, Dandy, Nikkie, and Spin looking on. Puist is sitting on the right, and on the wall Mama (left) is trying to get some leaves from her friend Gorilla. Later on Luit threw more leaves out of the tree, enough for everyone.

confrontation. Only Puist actually joins in the fighting, though unfortunately it is not clear whose side she is on. Almost immediately after they have all descended from the tree Yeroen bluffs over Luit. The two males then kiss each other and lick each other's wounds. Luit's wounds are clearly deeper.

The incident had one repercussion that is worth mentioning in connection with Mama's possible role as the leader of the female support for Yeroen. When the skirmish in the tree had come to an end and everyone was back on the ground, Mama made a bee-line for Jimmie, who hastily grabbed her child and fled from Mama, screaming loudly. At the same time Mama's friend and ally, Gorilla, attacked Krom. The reason for these two sudden attacks must have lain in the preceding conflict between the two males, but Jimmie and Krom had had nothing to do with the conflict. Perhaps their neutrality was Mama's reason? We shall never know for certain, but it was not the only time that the main conflict between Yeroen and Luit had repercussions among the females.

Day 64

The final serious fight takes place after Luit has been following Yeroen for a time and has pelted him with stones. Yeroen flees screaming, but then suddenly stops short and turns to fight Luit. Puist runs up to the fighting males and throws herself into the fray; once again it is not clear whose side she supports. It seems most likely that she is against Yeroen, because immediately after Puist joins in, Yeroen disentangles himself from the mêlée and flees screaming to Mama. Luit pursues him, but together with Mama Yeroen is able to fight him off.

When things have settled down again Puist comes over to Mama and "greets" her submissively. Yeroen and Luit do not reconcile their differences for a good half hour. This time Yeroen has been wounded but Luit has not.

We never saw wounds on either Yeroen or Luit other than after the two fights in the sleeping quarters and the three fights in the group. We can therefore safely assume that the two rivals had only five serious confrontations. The three fights in the group, which I have just described, show an obvious turnabout. In the first fight Yeroen was the attacker and winner, and his opponent did not bite back. In the second fight Luit did bite back. In the third and final fight Luit was the attacker and winner. In view of the fact that these three fights took place within a few weeks of each other an explanation cannot be sought in a change in the physical powers of the two males; the outcomes in fact reflect the change in the relative social status.

We tend to think that the outcome of a fight determines the social relationship, whereas here the outcome was determined by the social relationship. The same was seen in later dominance processes. The prevailing social climate affected the self-confidence of the rivals. It was as if their effectiveness depended on the attitude of the group (rather like a soccer team playing better at home than away). Although after four weeks Luit had shown in the fight in the sleeping quarters that he was physically stronger than Yeroen, it took him nine weeks to prove himself as convincingly in the group as a whole. By then Yeroen could hardly hope for any support, and one of the females, Puist, had probably deserted his camp altogether. Luit had carefully tested the reactions of the group before he dared make an overt attack on Yeroen. His success in the final fight was more than just a demonstration of brute strength: he made it quite clear to Yeroen that the attitude of the group had changed radically.

I cannot believe that fights between males are in fact tests of strength, because the males are too controlled. They only bite extremities, usually a finger or a foot, and less frequently a shoulder or the head. This special kind of controlled fighting is already obvious between two juvenile males, Wouter and Jonas, both in their play and during the rare serious incidents between them. Since this is virtually the only way the males fight each other, they cannot be out to prove their respective physical strengths. The crucial factor is their capacity to fight effectively within the rules. A male must be able to get his hands and feet quickly out of the way, and he must equally quickly be able to seize hold of his opponent's hand or foot. Speed and agility are just as important as strength.

The inhibitions and rules governing male-male confrontations in chimpanzees are characteristic of species with multi-male societies. This condition is far from common. Social mammals usually live in groups of several, sometimes many, females but few adult males. In some species, such as elephants, males are not really part of society at all, whereas in most other species a single male keeps rivals away from "his" females. Less commonly, males tolerate each other's presence, usually with little friendly contact among them, and the rarest situation of all is that males are actually friends and allies.

The other three great ape species are typical, ranging from intolerance among adult males to tense but rather uncooperative relationships. Male orangutans roam vast territories in the rain forest to ward off other males. Gorilla males sometimes live in the same group but typically monopolize females and fight off intruders to the death. Bonobo males do live together but are highly competitive: they do not hunt together as do male chimpanzees, nor do they defend a territory together or form political alliances. Bonobo males, including fully grown ones, follow

The outside enclosure at Arnhem Zoo is approximately two acres of forest surrounded by a moat. Two-thirds of the island is visible here.

To demonstrate our unique feeding experiment to zoo visitors and television crews, we arranged feeding sessions with reversed roles between human and ape. Instead of the ape in the cage and the human outside, here we have Monika protected by a cage trying to get Gorilla (who is holding Roosje) to pay attention to the procedures.

Mama, the group's matriarch.

At the age of eight weeks, Roosje was still in human hands.

Opposite: *Nikkie at the age of eighteen years.*

Yeroen, the oldest male, was the leader during the early years. Long thereafter, he remained a power to be reckoned with.

Yeroen (right) presents himself with spread legs to Franje, who ignores his advances.

Zwart (right) watches as Jonas fishes for food in the moat.

Wouter (left) *and Jonas, both five years old, tickling each other with playfaces.*

Opposite: *Young chimpanzees have a light-colored face that later often turns dark. Zwart with her one-year-old daughter, Zola.*

During the brief period of Luit's leadership, Nikkie (left) bows for him. Even though Luit, with his hair erect, looks larger, both males are approximately the same size.

Preceding pages: *Males usually use only their hands and feet when fighting females. Spin (right) protests at having been slapped by Luit. His slaps were so hard that she turned head over heels several times in the sand.*

A coalition of four aggressors united against a bluffing and side-stepping
Dandy. Left to right: Yeroen, Gorilla, Mama, Wouter, and Dandy.

Nikkie engages in a bluff
display with hooting. This
is the usual way social
climbers announce their
aspirations.

A mouth-to-mouth kiss between Nikkie (center, left) *and Hennie seals their reconciliation after an attack by Nikkie. Mama* (left) *and Dandy* (right) *look on.*

Reconciliations among adult males depend on who is the first to make a move. Nikkie holds out a hand to Luit about ten minutes after their conflict ended in the tree. Just after this photograph was taken, the males embraced each other and climbed down together.

Yeroen and Nikkie demonstrate their bond. Yeroen (front) *is mounted and embraced from behind by Nikkie, while both scream at their common rival, Luit, who is not in the picture.*

In the summer of 1997, I returned to the Arnhem colony to collect photographic updates on some of the book's main characters. Giambo's appearance came as a complete surprise. His hair color is a mixture of brown and grey, which makes him stand out among all other chimpanzees in the colony. His mother, Gorilla, is as black as a chimpanzee can be, and none of the potential fathers looks like him.

Opposite: *After a long and eventful life, Mama is old and looks it. Despite her rapidly deteriorating health, she remains the most influential female.*

Compare this photograph of Zola at the age of eighteen with the one taken when she was only one year old (page 7 in photo gallery). Now Zola is as dark as her mother, Zwart, whose name means "black."

Tepel (right) with her eleven-year-old daughter, Tesua.

their mothers around through the forest and depend on her for status: sons of high-ranking mothers tend to be at the top. Bonobo society is female-bonded and female-dominated, which is fascinating by itself but makes it a poor model for the complex male-male relations that characterize our own societies.

Male chimpanzees stand alone among our closest relatives in their ability to overcome the basic competitive tendencies found among all male animals and to achieve a high degree of cooperation. In the same way that many human males engage in continuous office rivalries while maintaining unity against corporate "enemies," chimpanzee males contain and ritualize their competitiveness because of the need to form a common front against their neighbors. Even if neighbor groups are absent in Arnhem, the chimpanzee male psyche, shaped by millions of years of intergroup warfare in the natural habitat, is one of both competition and compromise. Whatever the level of competition among them, males count on each other against the outside. No male ever knows when he will need his greatest foe. It is, of course, this mixture of camaraderie and rivalry among males that makes chimpanzee society so much more recognizable to us than the social structure of the other great apes.[8]

The Price of Peace

Because the five serious fights were so sensational, we perhaps tend to attach too much importance to them. In between these fights, literally hundreds of displays and nonviolent conflicts took place. The fights dramatically demonstrated the state of affairs, but this was ultimately determined by a complex series of factors that were reflected in innumerable and interminable social maneuvers.

Offset against the few serious outbreaks of aggression between Yeroen and Luit is the vast amount of time and energy they devoted to suppressing it: at least that is the most likely function of the many friendly contacts and long grooming sessions between the two rivals. Males tend to groom each other during periods of tension, a fact which is excellently illustrated on the graphs we plotted of the apes' grooming activities over the years. Grooming peaks between pairs of males coincided with periods during which the relationship between them was unstable. The record during one of the dominance processes was almost 20 percent; in other words, despite all their confrontations the two rivals spent about one-fifth of their time grooming each other. Yeroen and Luit did not groom each other as much as that, but their grooming activity was far higher than usual. Despite the fact that grooming appears to be a highly relaxed activity, it seems reasonable to assert that when males begin to groom each other more regularly it is a sign of friction. I emphasize the word "males," because it is not certain that this rule also applies to female chimpanzees.

A grooming session generally takes place shortly after a reconciliation. To begin with, the reconciliations between Yeroen and Luit baffled us because the two males would sit facing each other with their hair on end as if they were about to display again. The first afternoon, I watched this with a mixture of fright and amazement, but later I learned the difference between display behavior that leads to friendly contact and normal displays that aggravate the conflict. During reconciliation both individuals are always unarmed (no sticks or stones in their hands) and there is eye contact. Chimpanzee males avoid looking at each other in moments of tension, challenge, and intimidation. In moments of reconciliation, on the other hand, they look each other straight and deep in the eyes. After a conflict the former opponents may sometimes sit opposite each other for a quarter of an hour or more, trying to catch each other's eye. Once the opponents are finally looking at each other, first hesitantly but then more steadily, the reconciliation will not be far away.

Apart from the many scenes of reconciliation between Yeroen and Luit, we also witnessed truces. At least I can think of no better explanation for the long, intensive contacts between them just before they retired to their sleeping quarters in the evening. They never came into their cage unreconciled, and the fact that they had only two night fights may well be due to these truces. To illustrate this more clearly let me describe one of these evening "handshakes at the door."

Day 1: 5 P.M.

All the apes are inside except Yeroen and Luit, who are by the entrance. Luit is displaying slightly a little way away from Yeroen. Yeroen moves closer to Luit and pants at him. He holds out his hand and grins. Luit grins back but keeps his distance. With huge steps and hair on end Luit paces around Yeroen a few times, while Yeroen is still begging for contact. Luit takes a few steps toward Yeroen but then slowly backs away again. While he is backing still further away Luit keeps on grinning at Yeroen, and at the same time he has an erection. About 20 meters away from Yeroen, Luit lies down on his belly and makes pelvic thrusts into the sand, panting all the while. Yeroen approaches him hesitantly. Quite unexpectedly both apes begin to scream at exactly the same moment, and Luit presents himself to Yeroen, who begins to groom Luit's backside, panting heavily as he does so. After a few minutes the two of them sit down and proceed to groom each other for a long time. An hour later they wander into the building. The other apes have already eaten. The keeper and I have waited patiently until the two males had calmed down sufficiently to enter their sleeping quarters together. They both eat heartily.

These reconciliations and truces between Yeroen and Luit were not automatic. It was more that their "sense of honor" seemed to be at stake. If neither one was prepared to make the first conciliatory move—by looking at the other, holding out a hand, panting in a friendly way, or simply going up to his opponent—the two would continue to sit tensely opposite each other, and it was frequently a third party who helped them out of the impasse. This third party was always one of the adult females. Female mediation might take any number of forms. On the first day, for example, Krom was a party to the reconciliation I have described between Luit and Yeroen (she groomed them in turn), but her role was not yet decisive enough to be called a genuine attempt at mediation, of which there were many obvious instances later. After the conflict between the two rivals had died down, the female mediator would walk up to one of them and kiss him or groom him for a short time. After she had presented herself to him and he had inspected her genitals, she would walk slowly over to his opponent. The first male would follow her, sniffing at her vulva every now and again and without looking at his opponent. The male's decided interest in the mediating female's behind was unusual. In other situations an adult male would not follow a female who had just presented herself to him, especially if she did not have a sexual swelling. (In fact females with a sexual swelling have never been seen to act as mediators. This is understandable, because they would only be a source of further disagreement.)

The male probably followed the female mediator as a kind of excuse to approach his opponent while at the same time not having to look at him. It was obvious that the female did not just happen to walk over to the opponent; her mediation was a purposeful act. She would regularly look around to make sure the first male was still following her, and if he was not she would turn around and tug at his arm so as to make him follow. When she and the male reached his opponent, the female would sit down and both males would proceed to groom her, one on each side. When the female discreetly withdrew a few minutes later the two males would continue grooming as if nothing had happened, but now, of course, they would be grooming each other.

This surprising catalytic function of the females can only be interpreted as meaning that it was in their interest that peace should be restored. All the adult females performed the role of mediator at one time or another, but not all their methods were as subtle as in the incident described above. The most striking example I have ever seen was a more or less forced contact between Luit and Nikkie, after the two of them had been in conflict: Puist kept on jabbing Luit's side with her hand until he was sitting close to Nikkie and had to choose between jumping away or moving still closer to his opponent. He chose the latter.

Now that we have considered grooming, eye contact, truces, and mediation, the reasons for my intense interest in the whole subject of reconciliation may well be clear. I believe that the social significance of this behavior cannot be overemphasized. It almost certainly plays a crucial role as a constructive counterbalance to forces that threaten to disrupt the life of the group. It is amazing that so little research has been done into this aspect of behavior. In the 1960s and 1970s vast sums were spent on research into aggressive behavior among humans and animals, but virtually nothing has been done on its resolution.

I have kept one aspect of reconciliation until last because it is closely connected with the termination of the struggle between Yeroen and Luit. This is the phenomenon of joint displaying, by both parties, as an overture to reconciliation. It seems to have a very special significance as a test of the state of the dominance relationship. Luit had not been "greeting" Yeroen for weeks, so there was no doubt about the way things were developing. In the absence of such submissive behavior the emphasis came to rest on proving the opposite: dominance. Initially, only Yeroen adopted a dominant attitude at the end of a conflict. He did this before seeking reconciliation contact. When the two males approached each other, their hair on end, making themselves look as big as possible

Luit's attempts to reverse the dominance ritual made Yeroen extremely nervous. Yelping, he has walked away from his rival and now approaches Zwart in search of reassurance.

and looking each other straight in the eye, it was Yeroen who executed the bluff-over. In other words, he stepped over Luit, with his arm raised. Only then did the two males kiss and groom each other or seek other forms of contact.

After a few weeks of frequent bluff-overs by Yeroen an interim period dawned, during which the reconciliations hardly ever involved displays. Then, very gradually, Luit started to try bluff-overs himself. His attempts caused extreme tension in Yeroen. The first time Luit tried to reverse the roles, both males approached each other displaying and drawing themselves up to their full height, until they faced each other on two legs. Luit raised his arm to bluff over Yeroen, whereupon Yeroen ran away screaming and broke into a temper tantrum!

It was not until the 49th day that Luit succeeded in bluffing over Yeroen for the first time. This later became the standard procedure, but the reversal was a very gradual process. For a number of weeks Yeroen and Luit bluffed over each other. They both tried to escape the role of the subservient, dominated party and as a result sometimes put the process of reconciliation at risk. On one occasion Yeroen even exceeded all the unwritten rules governing such confrontations, by butting his head sharply against Luit's chest as the latter was stepping over him. Luit fled screaming in protest. On other occasions it was Luit who refused to be bluffed over. He backed away on his legs as Yeroen approached, whereupon Yeroen began to yelp and ran to the nearest member of the group, sought reassurance in a brief embrace, and then returned to Luit.

In the last week of the whole process Luit's increasingly dominant behavior was finally underlined by Yeroen's subservient behavior. Once again this took place within the context of reconciliation. We observed a number of new developments: attempts at contact at the end of a conflict were now only made on Yeroen's initiative. Luit regularly refused to have anything to do with Yeroen, which meant that Yeroen had to make several attempts before Luit was prepared to accept his overtures. At such moments Yeroen uttered sounds in Luit's direction that, although strongly reminiscent of "greeting," were too soft and mumbled to count as such. At the time I did not recognize the possible link between these three phenomena—the need for contact by the one party, refusal of contact by the other and the mumbled "greeting." It was not until I studied later processes that the explanation suddenly occurred to me: it was as if conciliatory overtures at this stage in a relationship could only be successful if coupled with this soft "greeting." In this final phase the losing party's need for contact is so great that the winning party can blackmail him! The winner refuses to have anything to do with the loser until he has received some respectful grunts.

I have so often seen how "greeting" has paved the way for reconciliation with an initially unwilling dominant group member, that I now firmly believe that for the contact-seeker it is a case of "bend or break." (If this interpretation sounds farfetched, it is largely because we humans too readily believe that such behavior calls for cool, calculated thinking, but I am not so sure this is always the case. The intelligence and social awareness required to exercise emotional pressure on others can be subconscious. After all, human children can sometimes be dictators in the family from a very early age, without being rationally aware of what they are doing.)

As a result of this blackmail the first vague "greetings" were directed at the new dominant male's back, because the latter would walk away and only stop and sit down when he heard the soft grunts. This phenomenon of "greeting" the opponent's back lasted between Yeroen and Luit until the 72nd day, when Yeroen uttered a series of clear grunts to Luit's face for the first time. I count this as the first real "greeting," and as such it marked the end of the dominance process.

From that time on the number of conflicts and displays between Luit and Yeroen decreased drastically. Within a week the group was enjoying uncommon peace. Two weeks later the former rivals were even playing with each other again, which they had not done for a good three months. And they were not the only ones. Nikkie, Dandy, and even Mama, Puist, and Jimmie frolicked around with playfaces. This was a very unusual sight, because adult females are seldom in such a playful mood.

A stable hierarchy is a guarantee of peace and harmony in the group. And the figures back me up on this: from 1976 through 1978 there were a total of thirty-seven serious fights between adult males, the majority of which (twenty-two) took place in periods when the individuals involved were not "greeting" each other. These "greetingless" periods together made up less than a quarter of the total. This means that the risk of violence, however small in general terms, was almost five times as high when formal dominance relations were disrupted.

These figures strengthen the assumption that "greetings" have a reassuring effect. They probably serve as a kind of insurance for the dominant male that his position is safe. A show of respect in the form of "greeting" on the part of the loser in the dominance process is the price he pays for a relaxed relationship with the winner. That was how Yeroen bought his peace from Luit at the end of 1976 and how the adult females bought their peace from Nikkie. With the establishment of the new order the group would almost certainly have entered a period of stability, had it not been for the fact that Yeroen's overthrow was far more than just a shift in the hierarchy. Yeroen's fall suddenly opened up brand-new coalition possibilities, and chimpanzees, like politicians, never fail to recognize and seize such opportunities when they present themselves.

Formation of the Triangle

The growth of a plant is so gradual that it cannot be seen with the naked eye. But we can tell whether a plant is growing because we have learned to recognize the signs, such as half-open buds and young leaves. When I started studying chimpanzees, however, I knew of no such indicators to tell me that certain social processes were under way. New situations develop so gradually that the change becomes apparent only when the process has nearly ended. This did not mean that I was never able to follow the transitional stages, but I had to do so in retrospect by referring back to my notes and diary entries. In fact it is these daily observations that form the basis for this book. In addition the students and I conducted systematic research into grooming behavior, playing, coalitions, and "greeting," and recorded other data suitable for processing by computer. The result is the quantitative picture to which scientists attach so much importance, because it gives them a firm basis on which to form conclusions. It is by studying all this material that we can, in retrospect, reconstruct changes in social relationships.

The major change which took place in the Arnhem colony was the formation of a "male club." This was a typical example of a very slow, gradual process, invisible to the naked eye. Data collected in the summer of 1976 show how rarely two or more males would sit together without female company. At that time the chances of seeing an all-male subgroup were 1 in 10. In subsequent summers the probability ratio increased to 1 in 4 in 1977 and 1978 and to 1 in 3 in 1979. Over the years the males clearly began to form a subgroup of their own, thus bringing about a segregation of the sexes. In this respect our colony came to resemble groups of chimpanzees in their natural state, where males form separate bands and spend a great deal of time in one another's company.

The male trio Yeroen–Luit–Nikkie was often to be seen in the center of the enclosure, while the females were spread about in small groups with their children. This did not mean that there was no contact between the males and the rest of the group—the males still played regularly with the children and groomed the females. But there was a definite change in the group's social activity. The three males took to walking, grooming, playing, and relaxing together, and their rivalries were also confined more obviously to their own little circle. Since Luit and Nikkie had joined together to dethrone Yeroen by driving a wedge between

The all-male group—from left to right, Dandy, Yeroen, Nikkie, and Luit—soundly asleep in the warm sand.

him and the females, the struggle for power had become a private affair. The trio formed, as it were, a political arena or a center of power. The only female who was at all successful in maintaining contact with the males was Puist. She was frequently in their company, played an important role when there were conflicts between them, and to some extent acted as a buffer between them and the other females. Any female who sought the males' company but remained "too long" ran the risk of being attacked by Puist.

Puist fulfilled the same function with respect to the fourth male, Dandy, who associated freely with the other males in the first few years but was later ostracized. She would sometimes underline his isolation by maneuvering him away from the immediate vicinity of the three "big bosses," or, if she did not succeed in doing this on her own, by setting one of the other males on to him. For some reason Dandy had failed to secure a niche for himself among the males, and his contribution to group life had seldom risen above that of the lower-ranking females. The explanation for this could be sought in physical strength. Dandy was certainly smaller than the other three males, but he had proved himself a match for the females in fights, and there was even a period

The triangle: left to right, *Yeroen, Luit, and Nikkie*

when the females, including Mama, "greeted" him submissively. Dandy was a physically mature adult male in all respects. The reason for his lack of influence in the group therefore had to be sought in the social developments. Circumstances favored the growth of a close-knit, fairly exclusive triangular relationship between the ex-leader Yeroen, the new leader Luit, and the up-and-coming Nikkie.

The foundations for this triangular relationship were laid in the autumn of 1976. Luit's successful takeover, which came about thanks in no small way to Nikkie, was accompanied by a change in Nikkie's attitude to the ex-leader. Several times a day Nikkie would be seen sitting somewhere on his own, his hair on end, hooting. His hoot gradually swelled until it ended in a loud screech. Then he would dart across the enclosure and thump heavily on the ground or against one of the metal doors. On other occasions he would leap almost 2 meters into the air and crash both feet loudly against the door. To begin with, his intimidation displays did not seem to be directed at anyone in particular, but later on they took place more and more frequently in Yeroen's vicinity. Finally he started hooting directly at Yeroen. He would sit down opposite him and swing large pieces of wood in the air. Sometimes he would move a stick backward and forward in front of Yeroen's face, so that Yeroen was forced to duck or step aside to avoid him.

The strange thing was that Yeroen did not protest. He did not scream loudly and launch a counterattack, as he had done against Luit. He tried to ignore Nikkie as much as possible. It was as if he did not see or hear his challenger, although Nikkie was standing right in front of him hooting defiantly. On a few occasions Yeroen turned to Luit and asked his help by grunting at him briefly and then nodding his head several times in Nikkie's direction. Luit would reply by bluffing Nikkie away. On other occasions Yeroen held out his hand to Nikkie and yelped, as if he was begging him to stop. The two of them never fought, and every one of Nikkie's lengthy displays was eventually followed by the two of them grooming each other. These grooming sessions were a form of reconciliation, but they lacked the intense embraces which had accompanied the reconciliations between Yeroen and Luit. Perhaps the reconciliations between Yeroen and Nikkie were lukewarm because the tension was never very great between them.

At the end of October we heard Yeroen "greet" Nikkie for the first time. This marked the end of a dominance process that had been without bloodshed and that had only very slightly affected the group as a whole. The hierarchical order was now: 1 Luit, 2 Nikkie, 3 Yeroen, followed by the rest. The original order had been: 1 Yeroen, 2 Luit, and then the others, including Nikkie. Nikkie had risen from nowhere to the position of second-in-command, where he was now firmly entrenched between the two previously senior males.

Why did Yeroen not resist Nikkie's challenge? I can think of many reasons, but none of them can be conclusively proven. At the end of his struggle with Luit, Yeroen seemed to be exhausted. The conflict may have affected him so severely that he no longer cared whether Nikkie also preceded him in dominance. What is more, what was at stake this time was only second place in the hierarchy.

It may be that in 1976 Yeroen did not want any problems with Nikkie for reasons of strategy. The rivalry between Luit and Nikkie increased daily, and any resistance Yeroen might have offered Nikkie would have worked to some extent in Luit's favor. Yeroen may already have felt there was some advantage to be gained from an increase in tension between the other two males, and he may well have foreseen that this favorable situation would develop still further if Nikkie were to secure the position of second-ranking male. I do not know whether chimpanzees are capable of such predictions, but the fact remains that Yeroen later benefited enormously from the confrontation between Luit and Nikkie. By subordinating himself to Nikkie, Yeroen degraded himself, but he also maneuvered himself into the key position in the triangle.

Luit's New Policy

In 1977 a group of secondary school children visited Arnhem to observe the chimpanzee colony for a couple of days. When I asked them who they thought was the leader, twenty-one of them said Luit, three said Nikkie, and two said Yeroen. The children knew nothing about our dominance criteria and I had purposely given the males new names so that my students and I could discuss the apes without fear of influencing the children in any way. According to the children Luit was the dominant male because he was more self-assured and bluffed more impres-

Yeroen (right) *ignores Nikkie's noisy hooting.*

sively than the other two, because the members of the group were most in awe of him, and because Luit was so huge. From their description Luit had obviously grown into the position of alpha male. Like Yeroen before him he constantly had his hair slightly on end, so that he appeared bigger and more powerful. He looked magnificent.

But it was not only Luit's outward appearance and the way he bluffed that had changed, he also had adopted a brand-new *policy*. (The word policy is used here to denote a consistent social behavior with a view to achieving a certain aim, quite apart from whether this behavior was determined by innate tendencies, by experience and foresight, or both. For example, a mother chimpanzee who defends her infant each time it is threatened or attacked also conducts a policy of a kind: a policy of protecting her offspring.) To begin with, Luit, assisted by Nikkie, followed a policy that led to Yeroen's dethronement. As soon as this particular power takeover was behind him, Luit's social attitude altered totally. His new policy seemed to be aimed at a completely different objective, namely to stabilize his newly acquired position. He changed his attitude toward the adult females, toward Yeroen, and toward Nikkie.

Luit's new attitude toward the females was obvious from his behavior whenever serious quarrels broke out. For example, on one occasion a quarrel between Mama and Spin got out of hand and ended in biting and fighting. Numerous apes rushed up to the two warring females and joined in the fray. A huge knot of fighting, screaming apes rolled around on the sand, until Luit leapt in and literally beat them apart. He did not choose sides in the conflict, like the others; instead, anyone who continued to fight received a blow from him. I had never seen him act so impressively before. This particular incident took place in September 1976, only a few weeks after he had become leader. On other occasions he put a stop to serious conflicts less heavy-handedly. When Mama and Puist were locked in a fight he put his hands between them and simply forced the two large females apart. He then stood between them until they had stopped screaming.

Besides such impartial interventions Luit also intervened on behalf of one or other party. Once again, however, his policy changed. Instead of a *winner-supporter,* he became a *loser-supporter.* The term "loser-supporter" is used to describe a third individual who intervenes in a conflict on the side of the party who would otherwise have lost; for example, if Nikkie attacked Amber, Luit would intervene to help Amber chase Nikkie away.

Without Luit's assistance, Amber would never have beaten Nikkie. If Luit's interventions were purely arbitrary, he would be found to support losers about 50 percent of the time and winners the other 50 percent of the time. In fact, after his rise to power Luit began to show solidarity with the weaker party. Before, he had supported losers 35 percent of the time, but after his elevation the figure increased to 69 percent. The contrast between these two figures reflects the dramatic change in Luit's attitude. A year later Luit's support for losers had increased still further to 87 percent.

It was not surprising that Luit as the alpha male should set himself up as the champion of peace and security and try to prevent conflicts escalating by supporting the losers. This form of behavior is referred to as the *control role* of the alpha male, and is found in many species of primates. Less is known about the possible importance of this role to the male himself. There are indications among macaques that the group leader's protective role and his strong ties with the females exclude other males from the central position in the group. This naturally furthers the stability of his leadership. Rivals who are not prepared to be frightened away meet with a great deal of resistance. One such example was described by Irwin Bernstein, who observed a change of leadership in a group of macaques. He concluded that "young males of superior fighting ability cannot usurp power without the support of a sizable portion of a group."

It is conceivable that there is a connection between the protection offered by a dominant group member in his control role and the support he receives in return when his position is threatened. In other words, an alpha male who fails to protect the females and children cannot expect help in repulsing potential rivals. This would suggest that the control role of the alpha male is not so much a favor as a duty: his position depends on it. If looked at in these terms Yeroen's fall could be explained by his inability to protect the others effectively against the aggression of Luit and Nikkie. Luit's behavior can be interpreted in the same light. To begin with, he demonstrated just how little the females could count on Yeroen's support by attacking or abusing them in Yeroen's presence. Later his attitude changed completely, and he himself adopted the role of protector.

About four months after the struggle for dominance between Yeroen and Luit had ended, the females started to support Luit. During the winter of 1976 the number of female interventions in conflicts between Yeroen and Luit were nine out of ten times in favor of Luit. The same was true of their interventions in conflicts between Luit and Nikkie. Luit's alpha position had taken on a broader basis. Only Gorilla refused to desert Yeroen. In the period when the others were changing camps, a great deal of tension arose between Gorilla and Puist. The reason for the almost daily fights between the two females was probably the emergence of Puist as by far the most important of Luit's supporters. In the autumn we sometimes saw Puist rush to Luit's aid in a conflict, but Mama would realize what Puist was going to do and nip any such initiatives in the bud by chasing Puist away. In the winter Mama was less prepared to interfere (she was pregnant), so that the way was clear for Puist and the others to give vent to their "true feelings." But developments did not stop there, because in time Mama also switched her allegiance to Luit. Now the new leader was able to call on this powerful and influential female when the other males placed him in difficult positions.

If Luit had been the leader of a group of macaques, the support of the females might have been sufficient in itself. But among chimpanzees the tendency toward coalition among males is so strong that a leader must always allow for the possibility of other males ganging up on him, if his group contains two or more males other than himself. While Yeroen was at the pinnacle of his power this problem did not present itself. Yeroen had built up his position on the support of the females; Luit, the other adult male, was forced to live slightly apart. He was often seen on his own, far away from the rest of the group. At that time the group's structure strongly resembled that of a small macaque colony, with one absolute leader as the magnificent focal point and potential rivals forced into peripheral positions. Yeroen's leadership only ended when Nikkie had grown up sufficiently and had developed into a possible coalition partner for Luit. The subsequent dominance process not only affected and altered the individual rank of some of the leading group members, but it also affected the prerequisites of stable leadership. The new alpha male, Luit had to contend with not just one but two rivals. There was no point in Luit's trying to ban both Yeroen and Nikkie to the periphery of group social life. That would have been tantamount to political suicide, because the two ostracized males would have joined forces against him. The only course left for Luit was to try to convert one of the two males to his cause; he chose Yeroen.

Nikkie chases Luit and threatens him with a heavy stone.

Luit's choice proves just how much friendships among chimpanzees are situation-linked. After all, Luit had banded together with Nikkie against Yeroen and the females. Now he turned everything upside down and sided with Yeroen and the females against Nikkie. Whereas previously Nikkie's attacks on the females had done much to further Luit's own cause and Luit had at times even supported Nikkie against the females, now Luit intervened between Nikkie and the females sometimes even before an incident had developed. When Nikkie raised his hair and approached a female, swaying slightly, ready to attack, Luit would act by placing himself beside or in front of the aggressor, so that Nikkie did not dare go any further. On other occasions Luit would actually attack Nikkie and slap and kick him until he ran away screaming. Luit's attitude to Nikkie had hardened, and the number of conflicts between them increased.

The greatest source of conflict between them, however, was not the females but contact with Yeroen. Each one sought the ex-leader's friendship and would not allow the other one to sit by him. Whenever the other two were relaxing anywhere near each other Nikkie would start hooting and displaying some distance away, and he would continue to do so for some minutes until Yeroen got up and walked demonstratively away from Luit. Yeroen offered little resistance, but he was often doubtful as to exactly where he stood. First of all he would walk away from Luit, but when Nikkie stopped displaying he would return and sit by Luit, which made Nikkie start displaying all over again. When this happened Luit would place himself between Yeroen and Nikkie or he would chase Nikkie away, thus defending his contact with Yeroen.

Later still Luit began to take a more active interest in contacts between Yeroen and Nikkie. Because of his position he was able to act resolutely and effectively, usually by attacking Nikkie. This in turn meant that Luit usually won the competition and formed the stronger tie with Yeroen. By the end of the winter Luit had succeeded in establishing a situation whereby he himself had far more contact with Yeroen than Nikkie. To this extent Luit's strategy had been completely successful. But there was one factor he could not control, and it was upon this factor that the stability of his position as alpha male depended, namely Yeroen's own attitude.

The Closed Yeroen-Nikkie Coalition

On a beautiful spring day in mid-April 1977 the colony was let outside again. Over the next few months I became more and more certain that Luit was firmly in command. Several times a day Nikkie's behavior toward him was extremely subservient: he would make himself as small as possible, his hair flat against his body, "greeting" him loudly. If Luit approached him directly, Nikkie would leap backward in the familiar froglike fashion known as bobbing. On one such occasion he was so unaware of his surroundings that he leapt backward into the moat (which is luckily very shallow at the edge). Luit certainly made a deep impression on him.

Yeroen also "greeted" Luit, but not in such an exaggerated way. To have done so would have been beneath his dignity and unfitting for his age. I have never seen Yeroen deeply submissive: he "greeted" Luit softly and if necessary went as far as to bow his head, but no further. Luit in turn broke any contacts between Yeroen and Nikkie with an air of great assurance. On one of the rare occasions when Yeroen did defend Nikkie against an attack by Luit he immediately followed this by "greeting" Luit and grooming him, as if he were apologizing for what he had done. Luit then proceeded to underline his position by chasing after Nikkie a second time, but in this instance Yeroen let things be. At other times Yeroen bluntly refused Nikkie's pleas for help: he turned his back on him and walked away when Nikkie screamed and held out his hand to him. It seemed as if Yeroen had finally decided to support Luit. He was often seen sitting near Luit, and sometimes he even displayed with him against Nikkie. What is more, there was a neat explanation for the collaboration between the two older males. They had known each other long before the colony had been set up, and it seemed likely that their historical ties had finally resulted in a coalition between them.

However logical this explanation may sound, it is still a weak one. Coalitions based on personal affinities should be relatively stable; mutual trust and sympathy do not appear or disappear overnight. And yet there had been no evidence of stability so far, at least not in the coalitions of the adult males. Did Luit end his collaboration with Nikkie because he had suddenly taken a dislike to him? And was the formation of a coalition between Luit and the females a question of sudden, mutual sympathy? If friendship is so flexible that it can be adapted to a situation at will, a better name for it would be opportunism. To begin with, there was always the possibility that the relationship between Yeroen and Luit would survive beyond the point when it ceased to serve a useful purpose, indicating that it rested on true friendship, but subsequent events were to disprove this idea. Even this old tie was insufficient to withstand the continual struggle for power.

In August the triangular relationship gradually began to change. Both Nikkie and Yeroen became less submissive toward Luit and resisted his interference more and more often. When the leader displayed at the two of them, they were no longer intimidated. Yeroen began to scream and furiously attacked Luit while Nikkie kept close to Yeroen with his hair on end, as if he were threatening Luit. Nikkie gradually became more successful in separating the other two males. Luit tried to stop him doing so, but if Nikkie persisted, Yeroen would walk away from Luit leaving Luit powerless to alter Yeroen's decision. In short, the balance seemed to be shifting in favor of a coalition between the beta and the gamma males, which represented an extremely serious threat to the alpha male. The ever-increasing unrest culminated six weeks later in a massive fight. In the weeks leading up to the fight, first Nikkie and then Yeroen stopped "greeting" Luit, and the two males gradually moved closer and closer to a real coalition—a coalition that in 1980, three whole years later, was still as strong as ever.

The initial incidents in these weeks involved the two senior males. About twice a day Luit displayed at Yeroen in an effort to impress him, but instead of Yeroen "greeting" him as he had been doing, he would protest violently. Screaming loudly he would chase after Luit, who calmly side-stepped him and continued displaying at him. More often than not the chase ended high in the dead oak trees, where Luit would leap about impressively, Tarzan fashion, from one tree to another, with all his hair on end. Yeroen attempted to keep up with Luit but never actually dared touch him when he came close. Yeroen's behavior changed instantaneously, however, if Nikkie appeared and displayed at the bottom of the tree or climbed a short way up toward them. Then Yeroen would pluck up courage, scream in a more aggressive tone, and try to grab Luit's feet. It was obvious from Luit's face that he was thoroughly scared in such situations. He bared his teeth when he saw Nikkie approaching and sometimes even screamed and fled to Mama. And yet in the first few weeks Nikkie did not undertake any actions of his own. He displayed at

The growing coalition against Luit: above, Luit (foreground) *makes a charging display past Yeroen, who bluffs back from a rise;* Opposite, *Yeroen changes his tactics and chases Luit into the tree screaming as he does so. Luit does not bare his teeth. Below, now Luit* (second from left) *does bare his teeth as Yeroen* (right) *screams to attract Nikkie's attention. (Nikkie is displaying somewhere to their right out of the picture, Puist is standing next to Luit, and Wouter is sitting next to Yeroen.)*

Yeroen's side and allowed himself to be embraced by him, but no more.

At the end of such incidents, when everyone was back on the ground, Nikkie would display at Yeroen for as long as it took Yeroen to "greet" him. The whole conflict, which often lasted as long as half an hour, began with Yeroen's resistance to a display by Luit and ended with Yeroen submissively accepting exactly the same behavior from Nikkie. In this way the dominance relationship within the coalition was confirmed to both males after each joint undertaking.

In time the emphasis shifted to direct confrontation between Nikkie and Luit. Yeroen became more and more self-assured and bluffed so often that it was almost as if he had regained his former position as leader. But at the same time he became increasingly content to leave

Nikkie to fight out his duel with Luit, although he made it absolutely clear whose side he was on. When Nikkie began to display Yeroen usually went and stood close behind him, wrapped his arms around his waist, pressed his lower belly against Nikkie's bottom and hooted gently along with him. This gesture is called "mounting," and it certainly originates from a sexual act. In this context, however, it had no sexual significance; it was meant as a *demonstration of unity.* Nikkie and Yeroen literally formed a closed front. The first time Nikkie and Yeroen stood united in front of him, Luit reacted by collapsing in a heap, screaming and rolling around in the grass, and beating the ground and his own head with his fists. His two opponents stood screaming in chorus with Luit for a short time, watching his tantrum, but then they walked away together, leaving an inconsolable leader for other members of the group to try to comfort. From that day on, the number of conflicts between Luit and Nikkie rapidly increased. The rising tension was also reflected in the numerous grooming sessions between the two and the intensity of their reconciliations. Even during these reconciliations Yeroen would rush up to Nikkie and Luit, briefly mount Nikkie, and so underline once more how the situation lay.

Nikkie, who a year earlier was half playfully hooting along with Luit and throwing sand at Yeroen, had now done a volte-face. What is more, his behavior was far more threatening and his influence had become much greater. He was now more than just a weight in the balance between the two others. The accent now was on the balance between himself and Luit, and the weights were the females and Yeroen.

The second change of leadership, unlike the first, was not brought about by a challenger who was physically stronger than the leader. I would say that Luit and Nikkie were about equally matched as far as physical strength was concerned. For that reason Nikkie only dared challenge and provoke Luit when Yeroen was nearby. When this was not the case he had to be very careful. On a day when the group had to stay inside because of torrential rain, Luit relentlessly beat up Nikkie and bit him in several places. Nikkie did not fight back but fled with a brief scream to Yeroen. It would have been very unwise of Nikkie to fight back, because all the females were in the hall with them and they would certainly have come to Luit's aid.

Outside, the influence of the females was less, because they were spread throughout the enclosure, but even Luit on his own was a for-

*Intensive reconciliations between Nikkie and Luit; top, the desire to start groom-
ing at the anus is so strong that they adopt a "69 position"; bottom, afterward
they change to a more relaxed position. (On both photographs: Nikkie left, Luit
right.)*

midable opponent for Nikkie. The tension between the two males was most obvious when they were both displaying at each other. Both Luit and Nikkie did their best to show not the slightest trace of uncertainty in each other's presence—Luit with his characteristic hard thumps on the ground and Nikkie hooting and throwing carefully aimed stones. When they were out of each other's sight, however, they showed definite signs of fear. This was a case of *genuine bluffing,* in the sense that each pretended to be braver and less frightened than he really was. During one of their confrontations, for example, I observed a remarkable series of signal disguises. After Luit and Nikkie had displayed in each other's proximity for over ten minutes a conflict broke out between them in which Luit was supported by Mama and Puist. Nikkie was driven into a tree, but a little later he began to hoot at the leader again while he was still perched in the tree. Luit was sitting at the bottom of the tree with his back to his challenger. When he heard the renewed sounds of provocation he bared his teeth but immediately put his hand to his mouth and pressed his lips together. I could not believe my eyes and quickly focused my binoculars on him. I saw the nervous grin appear on his face again and once more he used his fingers to press his lips together. The third time Luit finally succeeded in wiping the grin off his face; only then did he turn around. A little later he displayed at Nikkie as if nothing had happened, and with Mama's help he chased him back into the tree. Nikkie watched his opponents walk away. All of a sudden he turned his back and, when the others could not see him, a grin appeared on his face and he began to yelp softly. I could hear Nikkie because I was not far away, but the sound was so suppressed that Luit probably did not notice that his opponent was also having trouble concealing his emotions.

On October 14 the tension erupted into a full-scale fight. It was the first time that the Yeroen–Nikkie coalition really balled its fist and demonstrated just how shaky Luit's position was. The conflict started at noon, after Luit had been displaying around Yeroen for some time.

Phase 1

Yeroen begins by bluffing back at Luit, but he does not dare to attack him and starts yelping. Yeroen goes over to Nikkie and holds out his hand to him. In the meantime Nikkie has started moving around the open-air enclosure, his hair on end, displaying as he goes, and every now and then attacking females. Yeroen continues to beg Nikkie for his support and succeeds in chasing Luit up a tree. Luit bluffs back at Yeroen but flees screaming when Nikkie finally comes over to join Yeroen. Luit leaps over into another tree and slides down to the ground. His opponents follow him, and he scrambles into a single, tiny tree, but this proves to be a dead end—he cannot get down again without confronting the other two males. Luit begins to scream as I have never heard him scream before. He seems to be completely panic-stricken. Although I have my camera with me, I am too nervous to take photos. With Luit marooned in the tree top and his rivals seemingly intent on catching him, I feel certain that the outcome will be fatal.

Puist runs up and chases Nikkie away, but not for long. Nikkie soon returns to take up his former position opposite Luit. While Puist barks threateningly at both aggressors, Mama slowly moves over to the scene of the conflict followed by nearly all the other members of the group. Mama has her hair on end, and she sits down in the grass about 10 meters from the tree in which Luit is trapped. Nikkie hastily climbs down and walks over to Mama, puts his arm round her, and screams. Mama pushes his arm away. She goes and sits a short distance away. She stares at Yeroen. Yeroen stays where he is sitting in the tree and holds out his hand to Mama, yelping as he does so. She does not react. If Nikkie and Yeroen were hoping to win Mama over to their side in this way, they were wrong.

Phase 2

Nikkie and Yeroen put up a deafening scream, and both of them climb up toward where Luit is sitting. Luit has no option but to fight back, because he cannot escape. Nikkie and Yeroen grab hold of him and bite him, but this unequal struggle does not last long, because the highest-ranking females band together and quickly follow Luit's attackers up into the tree. Yeroen is bitten by both Mama and Puist. Mama then drags Yeroen out of the tree and chases him, screaming furiously, all over the enclosure. Puist stays in the tree and, together with Gorilla, launches an attack on Nikkie. Now that Yeroen has been driven away, Luit comes down out of the tree top and joins in the attack on Nikkie. Nikkie is finally defeated by an alliance between Luit, Gorilla, Puist, Oor, and Dandy. The rescue attempt is successful and has taken less than a minute.

Nikkie (left) looks larger than life, walking upright with bristling hair, while Luit avoids him with "greeting" sounds.

Phase 3

Everyone takes a breather. I have never seen so many chimpanzees wounded at one and the same time. Luit has an injured finger and foot, Nikkie an injured foot and back, Puist an injured foot, and Yeroen a scratch right across his nose. All the wounds are superficial, although Luit does not walk on his hand for days afterward. (Instead, he supports himself on his wrist. Amazingly, all the young apes imitate him and suddenly begin stumbling around on their wrists.)

Although his supporters have come to his rescue, Luit is clearly the loser. Nikkie behaves as if he has won. About five minutes after the conflict has ended Nikkie stalks over to Luit, his hair on end, and attempts a bluff-over. Luit refuses to bend down, but gives him a panting kiss instead and walks away. Half an hour later Nikkie goes over to Luit again and grooms him. Yeroen comes over to join them and starts grooming Luit as well. Peace has returned, but Luit's leadership is definitely at an end. The other two males have made him realize that their coalition is to be taken seriously, and there is nothing the females can do to change this. From that day on, Luit's position becomes progressively worse until, seven weeks later, he finally acknowledges Nikkie as his superior.

Nikkie's Absence

The dominance processes between the males in the colony created tension not only among the females but also among the human observers. At the time of Nikkie's rapid rise to power serious difficulties arose between the chimpanzees' keeper and myself. The keeper felt strongly that Nikkie was too young and uncontrolled to be the leader. In a way he was right, from what we know about chimpanzees in the wild, a male of Nikkie's age is perhaps not mature enough for a top position. My counterargument was that as leader he would be completely dependent on the oldest male and that we therefore did not have to fear an absolute dictatorship by a snotty-nosed upstart.

During the first weeks of his leadership Nikkie had a very hard time. By then it was winter again, and the colony was living indoors. In the comparatively restricted indoor conditions, it was inevitable that Yeroen and Luit should sometimes sit near each other. Nikkie tried to prevent this at all costs. Sometimes we would see Nikkie and Luit pushing and shoving to be near Yeroen, each trying to elbow the other out of the way. On one occasion this elbow work turned into a real fight (the only time I have seen males fight without displaying first).

One day, in this extremely explosive atmosphere, Nikkie picked up Wouter by the hand during a charging display and swung him furiously around above his head and against the wall. Although all the females sprang to poor little Wouter's rescue, screaming indignantly, and quickly succeeded in freeing him, the young chimpanzee limped for three weeks afterward and it was lucky that no bones were broken. It was obvious something would have to be done about Nikkie's rash behavior. The keeper and I decided to remove Nikkie from the group for the rest of the winter. Next summer he would be put back, and we would see how things went. The risks would not be as great in the outdoor enclosure. We expected one of two things to happen: either that the two senior

Mama screaming.

males would seize their chance in the interim period to strengthen their old ties so that Nikkie would never again succeed in prising them apart, or that there would be a struggle for dominance between Yeroen and Luit, which would probably be won by Luit, and hence we would be back to exactly the same situation as before Nikkie's bid for power. And what would there be to prevent Yeroen from joining forces with Nikkie again when he returned?

As soon as Nikkie was removed from the group the relationship between Yeroen and Luit changed drastically. Whereas in the preceding period Luit had constantly tried to sit next to Yeroen and had even got into conflicts with Nikkie for precisely that reason, he now kept his distance and avoided him. Their rivalry grew day by day, and sure enough a struggle for dominance flared up again between them, this time with Yeroen as the initiator. He emerged the loser, however, because all the females, except Gorilla, supported Luit. After a good two months of displays and conflicts Yeroen gave up. It was as if Yeroen's decision brought great relief to all the apes, because at his very first soft "greeting" the whole group rushed up to the two males, hooting, and embraced and kissed them both. There was general embracing, almost a dance of joy to celebrate the acceptance of Luit's leadership. This unusually strong reaction can be explained by the fact that it was the only dominance process to take place indoors, where there is no room to escape from the tension. The other members of the group seemed to know that Yeroen's first "greeting" represented an end to the whole process, and they were right. From that day, March 16, the relationship between Yeroen and Luit quickly improved. On the same day they groomed each other longer than they had ever groomed each other before, and after a few more days, with a lot of grooming contact, a period of real relaxation began, during which they even played together. Now we were anxious to see whether this new, friendly relationship would continue when Nikkie returned.

10 April 1978

The group is outside when Nikkie is reintroduced. The trapdoor is opened and . . . Nikkie does not dare to come out. Luit rushes inside and attacks him. Screaming, Nikkie flees out into the enclosure and rushes at top speed for the tall trees. The whole group chases him, *except Yeroen*. While Nikkie sits up in the top of the tree grinning with fear, the rest gather below. The first chimpanzees to climb up and make friendly gestures are Gorilla, Oor, and Dandy. From that moment on, Gorilla adopts the role of Nikkie's defender. She chases away Krom and Jimmie, who are still near Nikkie barking at him threateningly. A little later Puist attacks Nikkie, and Gorilla intervenes on Nikkie's behalf.

When Nikkie finally descends he is extremely frightened of Luit and turns and runs several times when he approaches. Finally Nikkie holds out his hand to Luit. Luit takes hold of Nikkie's hand and allows him to fondle his scrotum (a common form of reassurance among male chimpanzees). Afterward Yeroen joins them and gives Nikkie a kiss. Later that morning Luit and Nikkie have a long grooming session. When Yeroen attempts to groom Nikkie, Luit successfully intervenes and prevents contact between them. The same situation occurs again in the afternoon, but this time Luit does not succeed in separating them because, as he approaches, Yeroen and Nikkie react by screaming and embracing each other, and then together they chase Luit up a tree. Thus, on the very first day, Luit is confronted with the old coalition between Nikkie and Yeroen.

The next few days were marked by an atmosphere of conspiracy against Luit. While Luit sat surrounded by the females and children, the two contestants sat together, a little way off in another corner of the enclosure. Gorilla was the only one who had frequent contact with them, and we noticed how frequently she kissed Nikkie. Luit sometimes tried to stop her visiting his two rivals—he would place himself in front of her or display at her—but he never attacked her. During the previous winter Gorilla had reaffirmed her untiring support for Yeroen, so it was not surprising that she should extend her loyalty to his coalition partner. What did surprise me greatly was that this never gave rise to obvious problems between her and her friend, Mama. If a conflict broke out between the males, Mama and the other females would turn on Nikkie and chase him away, while Yeroen and Gorilla would defend him. In doing so Gorilla attacked any number of females, but it was as if she did not see Mama. The same was true in reverse; Mama undertook nothing against Gorilla, although her friend's behavior undermined her own efforts. Their friendship was obviously so strong that they were able to bear this important difference in attitudes. I have *never* witnessed a conflict between these two females.

Several times a day Nikkie would approach Luit, who was almost always somewhere near Mama. With his hair raised he would sit down opposite him and hoot defiantly. Yeroen would sit close up against Nikkie's back, hooting, with his mouth to his ear. If, after receiving encouragement in this way, Nikkie got up and began throwing sand and sticks, the females would mount an aggressive opposition, but Luit did no more than run along listlessly *behind* his supporters. He obviously did not rate his chances against the opposing threesome very highly.

Within a week Nikkie was the alpha male once more.

RESTLESS STABILITY

THERE IS MORE TO CHIMPANZEE SOCIETY THAN POWER TAKEOVERS. The picture I have drawn in this book so far is completely one-sided, in the sense that it only deals with the hard, opportunistic aspect of colony life. It is the impressive charging displays and noisy conflicts between the males that demand immediate attention. However, while the social hierarchy is stable, it is possible to look at a whole host of other, no less fascinating elements of chimpanzee life, such as the formation of social ties, the different ways in which females bring up their children, reassuring and conciliatory behavior, sexual intercourse, and adolescence. Each element represents another angle from which the life of the group as a whole can be studied. From some of these viewpoints the three principal males appear to be little more than extras. And it is difficult to say whether any one angle is more correct, more typical, or more important than another.

The classical lens through which Western scientists have tended to study the social behavior of animals has resulted in a sharp focus on competition, territoriality, and dominance. Ever since the discovery of the pecking order among hens by the Norwegian Schjelderup-Ebbe, in 1922, the status hierarchy has been seen as the main form of social organization. The study of monkeys and apes was for years dominated by attempts to rank individuals on a vertical scale, from high to low. There was, however, an exception: within the Japanese school of primatology, research workers were more interested in kinship and friendships. They classified individuals horizontally, representing them in a web of social connections. A distinction was made between central positions in the web and—in concentric circles around the heart of the group—increasingly peripheral positions. They were interested in the extent to which the other members of the group accepted an individual and which kinship group he or she belonged to. Broadly speaking, whereas the Western view sought to describe primate society in terms of a *ladder,* the Japanese were thinking in terms of a *network*. If we regard these two methods of approach as complementary it becomes clear why stable dominance relationships only partially ensure peace in the social system. "Horizontal" developments—in which children grow up and social ties are established, neglected, or broken—inevitably affect the temporarily fixed "vertical" component, the hierarchy.

One wonders whether Yeroen thought about the consequences of his cooperation with Nikkie before he embarked on the coalition.

The best indicator of a peaceful, relaxed atmosphere is play. Here the three adult males are playing together: left to right, *Nikkie, Luit, and Yeroen.*

This is one reason why hierarchical stability cannot be equated with stagnation and monotony. The second is that dominance must constantly be proven. An established hierarchy does not maintain itself. The Yeroen–Luit–Nikkie triangle was always teetering on the brink of renewed instability. The differences in dominance between the members of the triumvirate were slight, so that the balance of power might have shifted at any time. The possibilities were unchanged: on the one hand the choice between confrontation and reconciliation, on the other the patterns of coalition formation and isolation.

Below follows a brief description of the situation as it was in 1978–80, after Nikkie's power takeover. However tense the relationships may appear, it should always be remembered that aggressive outbursts were relatively rare. Separating interventions, displays, and "greetings" between the males were normal, daily occurrences, but entire days went by without any major conflicts, let alone real fights.

Divide and Rule

To begin with, Yeroen's behavior was somewhat confusing. Whereas he had at first reacted in an extremely positive way to Nikkie's return to the group, within a week his attitude had changed completely. He screamed in protest at Nikkie's displays. He was often successful in mobilizing the whole group against Nikkie and defended anyone who was attacked or threatened by him. In fact Yeroen was now undermining Nikkie's position, which Nikkie had secured largely thanks to his support.

During the first few weeks Luit fervently supported Yeroen in his actions, but after that he quickly lost interest, which is not surprising since he had little to gain from the situation. There was one thing Yeroen adamantly refused to tolerate, and that was actions against Nikkie which he himself had not instigated. On the rare occasions when Luit displayed at Nikkie or attacked him on his own initiative, Yeroen once more sided with Nikkie. Yeroen's fluctuating behavior excluded any possibility of Luit breaking through to the top again. But Luit did not allow himself to be "used" by Yeroen for long; his attitude toward Yeroen slowly changed from active supporter to neutral sympathizer.

The females' support for Yeroen also gradually decreased. But this was almost certainly due to Nikkie's systematic strategy of dealing separately with each group member who had opposed him in a conflict. Once he had reconciled his difference with Yeroen at the end of a conflict—even if he did not do this until half an hour afterward—he would then proceed to castigate the females who had sided with Yeroen. Even individuals who had done no more than bark at Nikkie from a distance were not spared. This policy of systematic reprisal sometimes resulted in a renewed alliance against Nikkie, but on the whole it must have worked as a deterrent. With the increasing neutrality of the females Yeroen found himself alone in his confrontation with Nikkie. What is more, Luit began to exert pressure of his own: when a conflict broke out between the other two he began to display. The only way to keep Luit in his place was to close the ranks. Since Yeroen on his own was no match for Nikkie's exuberant youthfulness, Yeroen was faced with the choice between submitting to Nikkie, or breaking their coalition, which would also certainly mean that Luit would return to power.

At the end of July 1978 Yeroen finally submitted to Nikkie, and from

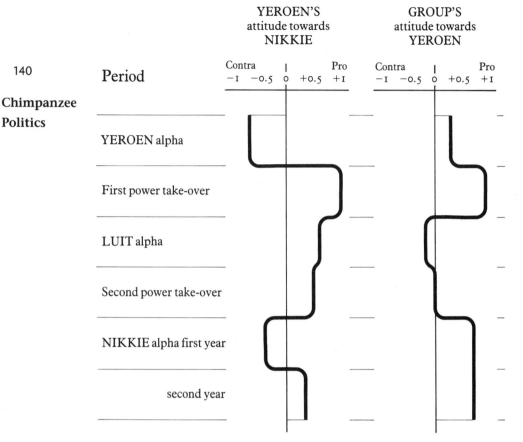

YEROEN'S attitude towards NIKKIE

GROUP'S attitude towards YEROEN

Period

Period
YEROEN alpha
First power take-over
LUIT alpha
Second power take-over
NIKKIE alpha first year
second year

Coalition Analysis

The way apes intervene in one another's conflicts indicates the existence of coalitions. The examples given here show Yeroen's attitude in conflicts involving Nikkie, and the females' and children's attitude in conflicts involving Yeroen.

In the first example, Yeroen's attitude was subject to great fluctuations. From the time his position as leader was threatened and usurped he supported Nikkie. He continued to support Nikkie even in confrontations with the new leader, Luit, thus making possible the latter's dethronement by Nikkie. Once this was a fact, Yeroen began to support Nikkie's opponents. After a short while, however, he reverted to his former role.

When Yeroen's alpha position was first challenged the whole group supported him. The second example shows how this support gradually dwindled, until the interventions in his favor were counterbalanced by as many interventions against him. Yeroen regained the general support of the females and children after the second power takeover. After that the group's attitude toward him was much more positive than toward the other two senior males, including the new leader.

that time on a strong link was forged between himself and Nikkie. Their coalition was to prove long-lasting, although in the following years conflicts still broke out between them from time to time. On such occasions Luit's threatening behavior forced them to break off and hastily reconcile their differences. Luit's displays and attacks caused chaos in the group, to which calm could only be restored by a joint action of the other two males.

During their coalition Yeroen and Nikkie did almost everything together. They sat and walked together, they bluffed side by side, and together they controlled Luit's isolation, by cutting short his contacts with high-ranking females and with Dandy. In all of this Yeroen encouraged Nikkie and acted as his adviser. For example, the three senior males are sitting in the shade and Dandy comes over and sits down by Luit. Yeroen is immediately aware of this and attracts Nikkie's attention by making a series of short grunts at him and, when Nikkie looks up, nodding his head in the direction of Dandy and Luit. Nikkie leaps up, mounts Yeroen briefly, and then chases Dandy away. Such incidents occurred regularly and gave us the impression that Yeroen had a sharper eye for potentially dangerous developments and realized better than his partner that such developments must be nipped in the bud. That Yeroen should be more vigilant was understandable in view of his age and experience.

It was almost impossible not to think of Yeroen as the brain and Nikkie as the brawn of the coalition between them. Yeroen gave the impression of a crafty fox, whereas with Nikkie his strength and speed were his most striking characteristics. And yet Nikkie's successful rise to power and the subtle policy he adopted afterward hardly fit the picture of a brainless muscleman. His policy with respect to the two males was one of divide and rule, which paralyzed both of them and made them dependent. If tension or even a conflict arose between Yeroen and Luit, Nikkie did not interfere, unless one of the two was obviously getting the upper hand. The displays by his great rival, Luit, were to some extent to Nikkie's advantage, because they forced Yeroen to seek protection. Sometimes Nikkie seemed to want to emphasize Yeroen's dependence, which he did by walking away as Yeroen came up in search of refuge. This meant that Yeroen had no choice but to follow his partner. The protection he got from Nikkie was minimal. Nikkie only intervened if Luit's aggressive behavior continued for any length of time and if he displayed very close to Yeroen. Nikkie's intervention usually took the form of his bluffing over Luit, who then ceased any attempts at intimidation. Yeroen often plucked up courage when Nikkie intervened and began himself to display at Luit. It was as if he were trying to exploit the change in the situation. But once again Nikkie might assert himself to stop such actions. So, while Nikkie might protect his coalition partner against their common rival, he did not allow him subsequently to turn things to his own advantage. Nikkie balanced one male against the other.

The relationship between Yeroen and Luit was very tense. After Nikkie became leader they never "greeted" each other. The lack of dominance of one over the other was also reflected in frequent *mutual* bluff displays. Such confrontations only took place when Nikkie was near at hand (if Nikkie was not present, Luit was too strong for Yeroen). They would approach each other, both of them with their hair on end, and neither would be prepared to step aside or to bend for a bluff-over. Sometimes they would even grab hold of each other, only to break loose again and run screaming to Nikkie. Yeroen would mount Nikkie from behind while Luit would approach him from the front to "greet" him. This would all happen in a flash. The characteristic outcome, with Nikkie in the middle and Yeroen "stuck" to him, suggested that the two older males did not dare to make a decision as far as their relationship was concerned without involving Nikkie. Presumably Yeroen was frightened of getting into a fight with Luit, and Luit feared an intervention by Nikkie. Nikkie acted as a kind of counterbalance. There were periods when this situation occurred several times a day.

The final aspect of Nikkie's policy was that he, naturally enough, did not tolerate the other two senior males being in each other's company. When he noticed they were sitting near each other, or that they were having some form of contact, he would display in front of them until they separated. Nikkie was extremely consistent in his interventions and almost always successful; in fact he *prohibited* contact between the other two males. It was clear that Yeroen and Luit knew this rule, because they only broke it with extreme circumspection. I once saw them grooming each other while Nikkie was asleep. They were able to continue undisturbed for a good five minutes—a stroke of luck—but they each, in turn, kept an eye on Nikkie; just like naughty boys who keep a lookout for the farmer when they have climbed into his orchard. As soon as Nikkie opened his eyes Luit strolled off, without looking back at the other two, as nonchalantly as possible so as not to attract Nikkie's attention.

Under pressure from Luit (left), Nikkie grins and holds out a hand to his coalition partner, Yeroen, with whom he has been in conflict. During conflicts between the other two males Luit carried out impressive displays, which Nikkie could only stop by mending the breach in his coalition with Yeroen.

The function of separating interventions is probably not only to prevent the formation of undesired coalitions but also to test existing coalitions. After all Nikkie would have been unable to drag the other two apart by force. He displayed at them from nearby and then waited to see whether they stopped having contact with each other. He was once observed displaying for a whole hour while the other two simply ignored him. Nikkie's prohibition could only be effective as long as Yeroen preferred a good relationship with him to having contact with Luit. Hence every successful intervention by Nikkie underlined the closeness of the existing coalition.

Contact between Yeroen and Luit occurred several times a day and was often instigated by Yeroen. Why did he do this? Would it not have been much easier for him to avoid Luit completely? After all, his relationship with Luit at other times was not particularly good, and when

the two of them did have contact they were disturbed, without exception, by Nikkie. My explanation is that this was Yeroen's way of making Nikkie feel dependent. Yeroen opened the door to a relationship with Nikkie's rival and was only prepared to close it again once his partner had felt the draft. It may be that Yeroen's behavior acted as a reminder to Nikkie, warning him that his position only rested on the way Yeroen chose to behave.

Collective Leadership

Nikkie had secured a strong position for himself. He was "greeted" by Yeroen, Luit, and the other group members and was thus the formal leader of the colony. But there was something lacking in his leadership. He met with great resistance from the females, and they found it difficult to defer to him. He was unpopular, and his authority was not readily accepted. He was "greeted," groomed, and obeyed, but not in the same matter-of-course way as the previous two leaders. He was feared rather than respected.

When Luit became leader he also became a loser supporter. He received the support of the females and rose so high in their estimation that they "greeted" him more often than Yeroen. Earlier on I explained these developments in the following way: a leader receives support and respect from the group in exchange for keeping order. The same phenomena occurred after the second power takeover, but the great difference this time was that these qualities were not seen in the leader. It was not Nikkie but his coalition partner, Yeroen, who defended the peace and consequently won general respect. This development surprised me more than anything else. Up to that point I had thought that this role, as well as formal dominance, had to be assumed by a single individual. Instead, whereas Yeroen and Luit had been sole leaders, Nikkie *shared* leadership with another individual.

The policing was done by Yeroen. Not counting the many times he and Nikkie intervened in each other's conflicts, Yeroen was a loser sup-

Opposite: *The leader will not tolerate his coalition partner's presence near his rival: top, Nikkie* (left) *goes and sits threateningly opposite Luit and Yeroen with his hair on end; middle, Yeroen walks away when Nikkie starts bluffing; below, Nikkie bluffs over Luit, and his separating intervention is successful.*

porter 82 percent of the time and Nikkie only 22 percent of the time (measured in 1978–79). Nikkie was still, despite his position as the alpha male, a winner supporter. Initially, following Nikkie's rise to the top, Yeroen worked against him so effectively that Nikkie could not really be said to be in control. For example, when the young leader, his hair on end, prepared to intervene in a conflict between two females or when he actually did intervene, Yeroen would immediately turn on him and chase him away, sometimes aided by the two females. It may well have been Yeroen's resistance to Nikkie's policing efforts, which in 1979 still prevented Nikkie from gaining complete control.

In Nikkie we had a leader who was regularly set on by alliances of females. What was even more surprising was that Yeroen encouraged the females in their resistance to his own coalition partner, although his encouragement became much less than it had been, with the result that the length and fierceness of these incidents decreased. The other side of the coin was that Yeroen was the male with the best political credentials as far as the females were concerned. The females turned against him for the duration of Luit's leadership, but after Luit's dethronement they returned to Yeroen's camp; they supported him more fervently than they did Luit, and in no sense did they support Nikkie.

The group's respect was accorded en masse to Yeroen. The females and children "greeted" him almost three times as often as Nikkie and five times as often as Luit. When Yeroen and Nikkie had ended one of their many joint displays and side by side approached a group of sitting apes, it was quite normal for these lower-ranking members to stand up and hasten to "greet" and kiss Yeroen while seeming to ignore Nikkie's presence. However, as the graph on page 148 indicates, Nikkie's scores eventually began to rise. In the course of 1980, Nikkie started to receive the same number of "greetings" as Yeroen, two full years after he rose to dominance.

Sometimes it seemed that Nikkie was being used as a figurehead, and that Yeroen—experienced as he was and extremely cunning—had him in the palm of his hand. The broad basis for leadership rested not under Nikkie but under Yeroen. The older male had a coalition with the females to pressurize Nikkie and a coalition with Nikkie to keep Luit in check. Seen in these terms the situation appeared to represent a comeback for Yeroen. Luit had deprived him of the support and respect he had hitherto enjoyed, but by pushing a youngster forward Yeroen seemed to have succeeded in reacquiring both.

A typical situation in the triangle. Nikkie (center) grooms his partner, while Luit sits alone a short distance away.

Luit cannot cope with the joint actions of Yeroen and Nikkie. Here he is helplessly throwing sand at them after having been chased away because he was grooming Mama.

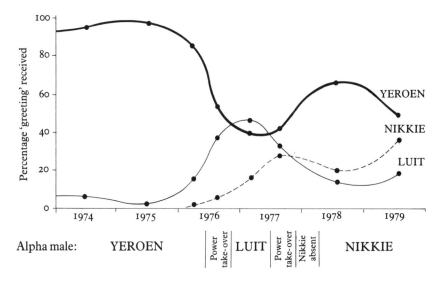

Respect Graph

*Respect for a high-ranking group member is measured, among other things,
by the frequency with which he is "greeted." The graph, made up from many
thousands of observations, shows how the females and children meted out their
respect to the three senior males in the years 1974-79. The total number of "greetings" per period is expressed as 100. "Greetings" between the males themselves
have not been included, although these constitute the criteria for the alpha rank:
the alpha male is the male who is "greeted" by all other adult males.*

*In 1974 and 1975 Yeroen received nearly 100 percent of the "greetings." During
1976, when Luit was challenging him, Yeroen's score dropped. During the period
of Luit's leadership Yeroen was no longer the most frequently "greeted" male,
but when Luit fell from power, respect for Yeroen grew again. It was not the new
leader, Nikkie, but his older coalition partner, Yeroen, who profited most from
Luit's dethronement. Over the years the number of "greetings" directed at Nikkie
steadily increased, so that by 1980-81 he had acquired the respect normally due
to the alpha male.*

This picture was not altogether correct. Yeroen had had to sacrifice
much for his "comeback." It was true that Nikkie did not dominate him
at all times, but he was strong enough for Yeroen to "greet" him. If
Yeroen refused to recognize Nikkie's position—and Yeroen certainly did
refuse in the first few months of Nikkie's leadership—this gave rise to
severe conflicts between them, and the coalition was in serious danger
of collapsing. Nikkie was dependent on Yeroen, but the converse was

Yeroen was not only "greeted" most often, he was also the one who needed to do the least to earn this "greeting." Even when he lay sleeping, females would come up to him and spontaneously demonstrate their respect. Here Oor (right).

equally true. Furthermore, Nikkie enjoyed the sexual privileges due to his rank, as we shall see in the next chapter. Nikkie occupied the top position, and Yeroen fulfilled the control role and had the authority that goes with it.

Nikkie's position was not an easy one. Compared to him Yeroen and Luit had been almost all-powerful, thanks to the collaboration of the females. The important difference between Nikkie's leadership and the old order was that Nikkie stood on the shoulders of someone who was himself very ambitious. The ensuing problems are familiar enough in the human world. Machiavelli wrote about the relative powerlessness of this kind of leader. If in the following quotation from *The Prince* we translate "nobility" as "males of high rank" and "common people" as "females and children," then we see that Nikkie's "principality" is indeed very different from the "principality" of his two predecessors:

> He who attains the principality with the aid of the nobility maintains it with more difficulty than he who becomes prince with the assistance of the common people, for he finds himself a prince amidst many who feel themselves to be his equals, and because of this he can neither govern nor manage them as he might wish.

SEXUAL PRIVILEGES

SOME VISITORS TO THE ZOO TURN AWAY SHOCKED, DRAGGING THEIR children with them, when they see the chimpanzees having sex. Others burst out laughing and make suggestive comparisons, and yet another group watches the scene in tense silence. Sex leaves no one cold. The swollen vaginal lips of the females immediately attract attention. Although outsiders may find it hard to believe, we are so used to the swollen pink bottoms of the females that we no longer think them grotesque, and on some of the females, Amber and Gorilla for example, we even find them beautiful and elegant. The public, however, finds them all equally disgusting and usually takes them to be chronic sores. On one occasion a woman introduced herself at the zoo entrance for the sole purpose of warning us that we had an ape with a monstrous red *head*. No doubt one of the females had spent some time standing on her head that day with her swelling pointing triumphantly in the air. This is quite normal behavior for an estrus female.

For a long time the traditional image of apes was of lustful satyrs living in constant perversity and sin, comparable with the people depicted in Hieronymus Bosch's painting "The Garden of Delights." It is not for nothing that the former Latin name for the chimpanzee was *Pan satyrus*. This man-ape from the jungle even had a reputation for raping human women. This age-old theme was again exploited in the film *King Kong*, which featured a gorilla. These stories of kidnapping and rape are nothing more than horror fiction; only apes that have grown up among humans show any sexual interest in them. And apes among their own kind are far from uninhibited. Their sexual intercourse is subject to clearly defined rules. Chimpanzees do not know the exclusiveness of pair formation, but neither is their sex life completely promiscuous.

The following figures show that our chimpanzees do not live a life of uncontrolled sexual activity. Chimpanzee females have an average menstrual cycle of thirty-five days, and their genitals are fully distended for about fourteen days. When the females are in this attractive phase of their cycle, the mating frequently per adult male is an average of once in five hours. This means six matings per eight-hour day for the female, because our colony has four males. This is the mating frequency for adult females, except Puist (who refuses to mate); in the case of adolescent females the mating frequency is more than one and a half times as high.

Nikkie invites a female to mate with him.

This is not because they are more attractive; on the contrary, the males are most interested in the mature females. It is the rivalry between the males for contact with the adult females that restricts the frequency of sexual intercourse with them.

Mating only occurs as frequently as described above when the female is in estrus. As soon as her swelling decreases the males lose interest. There are also long periods when the female's cycle stops or is very irregular (during the term of pregnancy of seven and a half months, and the nursing period of about three years). In a colony where a large number of infants are born, as in ours, this means that months can go by without any sexual intercourse at all between adult apes.

It is untrue to say that the life of the group is dominated by sex, but this does not mean that sex is unimportant. The adult males, for example, may refuse to eat for days on end when one of the females is in her estrus period. When I see them early in the morning, in their sleeping quarters, I can read the excitement in their eyes. They have a covetous look, which they also have when they get something especially tasty to eat. It is clear that they are anticipating the pleasures of the day ahead.

Males are very interested in the genital swellings of females. Here Luit inspects Mama.

Three males follow an exciting female, taking care not to lose sight of one another for a moment. The alpha male, Nikkie, is recognizable by his hair, which is standing on end, and the female by her swelling.

Courtship and Copulation

Courtship between adult chimpanzees is almost exclusively initiated by the male. He places himself at a little distance, varying from 1 to 20 meters, from the estrus female. He sits down with his back straight and his legs wide apart so that his erection is clearly visible. His long, thin penis is pink in color and is therefore clearly distinguishable against his dark hair. Sometimes he flicks his penis quickly up and down, a movement that makes it all the more obvious. During this show of his manhood the male supports himself with his hands behind him on the ground and thrusts his pelvis forward. If the female is sitting with her back to him, he attracts her attention with a series of soft grunts. The deaf female, Krom, does not react to this signal, so the male throws pebbles at her and stamps his foot on the ground or on the branch of the tree where she is sitting. It is by no means certain that the female

Luit (left) *places his hand on Oor's back, after which she will crouch down ready for copulation.*

will even favor the male with a look, but if she does look his way, the male immediately raises his arm and holds it stretched out in front of him invitingly. If the female accepts his invitation, she crouches down under his arm, with her swelling between his legs. The male takes hold of the female's shoulder and carefully maneuvers his penis into position inside her. Copulation itself lasts less than a quarter of a minute and consists of a few deep, powerful thrusts. During this time the female lies motionless on her belly. Usually the faces of the love-makers are virtually expressionless, but young females sometimes emit a high-pitched shriek during climax. Only very rarely does the female turn her head so that the two are facing each other.

No remarkable deviations from the standard pattern were observed in our colony until Amber and Oor reached puberty. They demonstrated such decided partner preferences that one might almost call it infatuation, and they were so insatiable that they exhausted their partners. Amber was greatly attracted to Nikkie. When they groomed each other they cuddled up close, and when they indulged in sexual games they did so away from the group. In this way they avoided being disturbed by the chimpanzee children, who are fascinated by sex, or by Oor. Oor would come rushing up to Amber on two legs, thrashing threateningly with her arms, intent on interrupting her mating session with Nikkie. At other times she would quickly present herself to Nikkie just as Amber was about to do so. Her interference only decreased after she had developed an intimate relationship of her own, with Dandy. These two young couples proved with their fondling, cuddling, and enthusiasm that chimpanzees are capable of playing with sex. This was clearest of all in the so-called "sex dances."

In a typical sequence Amber nudges Nikkie and together they find a quiet spot. Once there, Nikkie invites her to mate, but Amber crouches too briefly for copulation to take place. She skips off and starts bobbing at him with her lips pouted a few meters away. Sometimes she rushes up to Nikkie and presents herself, only to bound away again. Then she stretches up in front of him on two legs, scratches herself with long strokes, and at the same time takes a few paces forward. This pattern of approach, mounting, moving, half-skipping away, looks a bit like a dance, and even more so when Nikkie joins in by doing a brief gallop

Oor screams at the climax of copulation with Nikkie.

interspersed with a few wild leaps. The pattern may be repeated as many as fifteen times and ends with copulation.

The initiative among adolescents lies more with the female. She wants everything she can get and demands so much of the male that sometimes he cannot satisfy her. When he has reached this point he may briefly place a finger in the vagina of the presenting female. Usually he will avoid her. Both Amber and Oor seemed to have difficulty in accepting the limits of their partners' potency. If a young female has invited her partner to mate but been refused, she may well return to him after a short time, push his legs apart, and carefully feel his penis. Sometimes it is flaccid, but usually there is nothing visible at all, because chimpanzees can sheath their penises. If this insistent fondling of his genitals is repeated too often, the male gets fed up and moves away from his mate. The female may then throw herself down on the ground screaming in despair and have a tantrum, or she may rush after him wailing and yelping until he calms her by mounting her briefly (without an erection).

Amber's and Oor's preferences for Nikkie and Dandy respectively were restricted to their estrus periods. Their relationships were therefore of a sexual nature, but they were not true pair formations, because they did not exclude other contacts. The two males mated regularly with other females too, while the great receptivity of the two young females extended to other males and male children. But they only undertook initiatives of their own with their preferred mates, and they reserved their "sex dances" for them. In recent years these phenomena have been declining slowly. Perhaps this form of sexual enthusiasm is characteristic of the young and later only reflected in a mild preference. Among the older apes there are obvious sexual partner preferences, both in our group and among wild chimpanzees. This is one reason for not regarding chimpanzees as totally promiscuous. Another is the regulatory influence of the male hierarchy on sexual activities.

One of the students working for me, Mariëtte van der Weel, studied partner preference and the blossoming of Amber's sexuality in 1977. Mariëtte also studied a strange form of behavior among infants and juveniles, which is known in primatological literature as *sexual harassment*. When adults start a mating session the young come rushing up. They jump on the female's back so as to be able to push her partner away or touch him, or they wriggle between the couple. They also throw sand at them or, despite their size, conduct intimidation displays. Open

Amber's sex dance in front of Nikkie, which ends in copulation.

aggression toward the pair is extremely rare. The worst instance I have ever seen was when Fons bit Nikkie's testicles while Nikkie was mounting Fons's mother, Franje. This brought the session to an abrupt halt. But by and large these interferences are not hostile, and sometimes they look positively friendly. They are, however, clearly disruptive. If we consider that the children harass half of all copulating couples and that a good quarter of their interventions result in copulation being broken off, it is not surprising that the males often half playfully chase the children away before they make advances to an estrus female. But the children are like tiresome flies: they come back again and again. The little ones seem to be magnetically attracted to sexual contacts among their elders.

Why should this be? It is easy to give a psychological explanation: the children are simply jealous. However convincing this may sound, something is missing. I do not deny that they are jealous because in fact chimpanzees seem very jealous creatures, but there must be a purpose behind this sexual harassment, otherwise their social life would be tense and full of conflicts unnecessarily. Since Darwin, biologists believe in functionality. The anatomy, physiology, outward appearance, and behavior of animals have not evolved without a reason. A characteristic in which the negative aspects outweigh the positive aspects will not be handed down from one generation to the next. What then is the advantage of sexual harassment? One theory is that young chimpanzees try to prevent their mother from becoming pregnant again too soon, in order to postpone the arrival of their next brother or sister. If they succeed they will be able to benefit longer from their mother's milk, transport, and care. The children do not know why they act in this way. It is assumed that sexual harassment is an innate reaction that lengthens the infants' suckling period, hence their chances of survival.

When Fons sank his teeth into Nikkie's scrotum I expected Nikkie to turn on him in fury, but he did not do so. He rubbed the sore spot and looked at Fons, but he did not punish him. Chimpanzees are incredibly tolerant of infants. This is perhaps partly because aggression toward them tends to have a boomerang effect. As soon as the male threatens the infants harassing him and his mate, the female turns on him—even if she is in the middle of copulation—screaming in protest. No doubt he will have to forgo her sexual favors for some time after such an incident.

In the same way that the protective reaction of the female decreases as the children get older, so the male's tolerance toward them decreases.

While Franje mates with Nikkie, Fons, her son, comes over and embraces them both and gives Nikkie a kiss. Wouter (left) reacts by jumping excitedly around them and hooting.

This process begins when the children reach about four. Whereas before that age the males will tickle the children and chase them away half playfully, with older children they adopt a more authoritative attitude; the male barks threateningly at the juveniles and then expects them to stay away from the attractive female without any more ado. If they do not immediately obey they can expect to be taught a lesson. The adult male may bite their hand or foot and sometimes even draw blood. The fierceness of the punishment and the fighting technique used (which is typical of male fights) suggest that the juveniles are no longer merely regarded as "nuisances" but as potential rivals. All the older children in our colony are males and sexually active, although they have not yet reached sexual maturity. It is only in the sexual context that they are

The males become sexually active at a young age. Amber "mates" with Wouter, whereupon Fons (left) reacts by bluffing. Between adult males this would lead to a serious conflict, but here it is still partially a game.

treated so roughly. In this way, they learn the harsh rules of the adult male world early in life. It is clear that the oldest children have learned their lesson and will not dare approach estrus females without due caution.

When these young males reach puberty in a few years' time, we will be faced with the problem of inbreeding. Sons will be able to mate with mothers and later on, when the females mature, brothers with sisters, and fathers with daughters. We do not know yet what steps we will have to take. The problem may not in fact turn out to be very serious. There are strong indications that chimpanzees avoid incest of their own accord. While some anthropologists regard the human incest taboo purely as a cultural product and even as a "most significant advance" upon animal behavior, biologists tend to think of it as a law of nature that has permeated all cultures. In 1980 Anne Pusey published some significant data about the wild chimpanzees in Gombe Stream. It appears that sexual activity among siblings is very low there, and mating between mother and son has never been observed. Young females are strongly attracted to unfamiliar males, whom they seek outside their own community. After mating they either return, pregnant, to their own community, or they stay with the new community. Females are cautious in accepting partners within their own group. Anne Pusey writes, "Four females frequently retreated, screaming, from the sexual advances of males in their natal group old enough to be their fathers, while they responded readily to the courtship of younger males during the same period by presenting and mating." The young females cannot know who their father is, but they avoid fertilization by possible fathers by refusing to mate with males who are both old and familiar.

Tarzan has reached the age when he has to learn the hard way who has sexual privileges and who does not. Nikkie has his foot between his teeth and swings him around and around.

Amber and Oor's attraction to the two youngest adult males in our colony fits this pattern, but the real test of the incest avoidance mechanism will be in a few years' time. At the moment, the juvenile males "mate" with every female who allows them to, even their own mothers. However, one of the mothers, Tepel, does not tolerate it. When Tepel is in estrus she refuses to "mate" with her sons, Wouter and Tarzan. She pushes them away as soon as they have an erection, but she allows other infants to try. I am curious to see whether the difference in upbringing between the various mothers will be reflected in the later mother-son relationships.

Ambition and Fatherhood

The animal world is rife with sexual rivalry between males. Even the sweet-sounding song of the male nightingale is an example of this bitter struggle. His song warns other males to keep out of his territory and attracts females. The formation of territories is one way of demarcating procreational rights; the formation of a hierarchy is another. There is a definite link between power and sex; no social organization can be properly understood without knowledge of the sexual rules and the way the progeny are cared for. Even the proverbial cornerstone of our society, the family, is essentially a sexual and reproductive unit. Sigmund Freud, speculating about the history of this unit, imagined a "primal horde," in which our forefathers obeyed one great chief, who jealously guarded all sexual rights and privileges for himself. This jealous but also charismatic man, the Father, was finally killed by his own sons and cut in pieces. Later a new form of group life grew up, again with a man at the top, but this new group was only a shadow of the former one, in that "there were numbers of fathers and each one was limited by the rights of the others." According to Freud we have never been able to erase this image of the almighty father figure completely, and he lives on in our taboos and religions.

When I am observing the Arnhem chimpanzees I sometimes feel I am studying Freud's primal horde; as if a time machine has taken me back to prehistoric times, so that I can observe the village life of our ancestors. They still accept the *droit du seigneur*,[9] one of the forgotten products of Western culture. When Yeroen was the alpha male he alone was responsible for about three-quarters of all matings. Not counting sexual intercourse with young females (who arouse less rivalry), his share was almost 100 percent. Sex was his monopoly in the group. This situation came to an end when Luit and Nikkie revolted against him. Yeroen was not cut in pieces, but he has never been able to regain anything like his former share of the sexual activity. What is more, no other male has been strong enough to monopolize estrus females as completely as he did in his prime. On the other hand, the matings under Nikkie's leadership were still not equally shared out among the four males in the group. In the period immediately after Yeroen's fall, it was Nikkie and above all Luit who had sexual intercourse. Yeroen's share only began to increase again when Luit was being dethroned (by Yeroen and Nikkie) and reached a new peak during the first year of Nikkie's leadership; at that time Yeroen scored higher than his coalition partner. In the year that followed, Yeroen had to take a step back again; now Nikkie's share was over 50 percent. In all these years Dandy's share was always under 25 percent, except in the confused months of the second power takeover.

There is, generally speaking, a definite link between the rank of a male and his copulation frequency, although it is by no means a rigid law but rather a rule to which exceptions are possible. It is not that high-ranking males are more virile, but that they are incredibly intolerant and chase lower-ranking rivals away from estrus females. If they catch another male mating, they intervene by attacking either him or his mate. The females are also clearly aware of this risk. Sometimes a female consistently refuses to accept invitations to mate from certain males, as if she is just not interested in them. Then, when the colony goes indoors in the evening, opportunities suddenly present themselves for undisturbed mating, and it turns out that the female is perfectly willing to mate with males she has cold-shouldered during the day. We have even seen females rush to the cages of males to copulate quickly through the bars. This only happens, of course, when the alpha male is still outside or separated from them in another part of the system of passages. If the alpha male happens to spot what is going on, he immediately reacts by hooting and bluffing, but he is powerless to intervene.

What is the reason for this intolerance? Why are the males unable to leave each other alone? Jealousy is once again only half the story. The problem of its function remains. Jealousy would have disappeared from the earth a long time ago if the tensions and risks involved did not have some positive function. The biological explanation for sexual rivalry

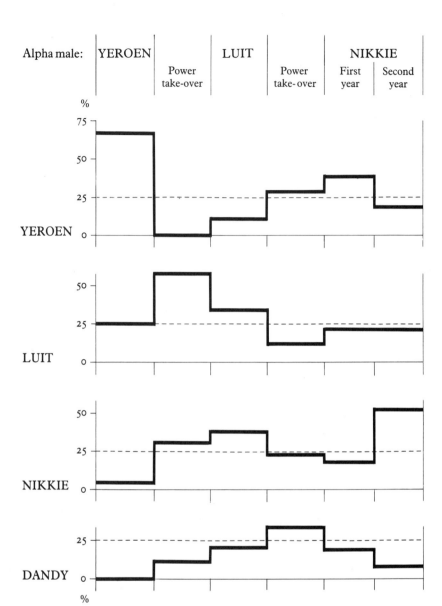

Alpha male:	YEROEN		LUIT		NIKKIE	
		Power take-over		Power take-over	First year	Second year

Mating Activity

If the total number of matings had been equally divided among the four adult males, they would each have accounted for 25 percent. In reality, however, during the years 1974–79 there were three periods when one male was responsible for over half of all matings: first Yeroen, then Luit during his bid for power, then Nikkie during his second year as leader.

between males is as follows. A female can only be fertilized by one male. By keeping other males away from her a male increases the certainty that he will be the father of her child. Consequently, children will more often be sired by jealous than by tolerant males. If jealousy is hereditary—and that is what this theory assumes—more and more children will be born with this characteristic, and later they in turn will attempt to exclude other members of the same sex from the reproductive act.

Whereas the males fight for the right to fertilize as many females as possible, the situation for the female is completely different. Whether she copulates with one or one hundred males, it will not alter the number of children she will give birth to. Jealousy among females is therefore less marked. Female competition occurs almost exclusively in pair-bonded species, such as many birds and a few mammals. In those cases, females try to gain or defend a long-term tie with a male. Our own species is a good example: research by David Buss has demonstrated that whereas men get most upset at the thought of their wife or girlfriend having sex with another man, women dislike most the thought that their husband or boyfriend actually *loves* another woman, regardless of whether or not sex occurred. Because women look at these things from the perspective of relationships, they are more concerned about a possible emotional tie between their mate and another woman.

Males focus on sex and power. Their drive for power derives from the fact that the male hierarchy determines sexual priorities. If striving for a higher rank translates into more progeny for a male, more sons will be born with this particular tendency. This theory explaining the origins of male ambition is simple, logical, and therefore compelling. To prove it, however, we would have to carry out a great deal of research. For example, it is important to know which mating sessions lead to fertilization and what positions in the hierarchy a male has occupied throughout his career, from very young to his death. The connection between rank and sex among baboons, macaques, and wild chimpanzees has been the subject of intensive study in recent years, and the evidence supporting the theory is reasonably strong, though not overwhelming. For example, the actual act of procreation is only half the job; children also have to be protected after they are born. Dominant males are in a better position to offer protection to the mother and her child. It is difficult to say whether this is an alternative to the old theory or an extension of it, but below follow several examples of the attitude of males in the Arnhem colony to infants.

Example 1

One day Jakie, who is less than a month old, is taken from his mother against her will by his "aunt" Krom. His mother, Jimmie, follows yelping and whimpering, but Krom refuses to give Jakie back—until, that is, Yeroen and Luit see what is going on, approach the two females, plant themselves threateningly in front of Krom, and display at her. Krom hastily returns Jakie to his mother.

Example 2

In another, similar incident Tarzan is kidnapped by his "aunt" Puist. Tarzan is about one year old and is riding on Puist's back. All of a sudden Puist climbs up a tree with the infant clinging on for dear life. When Puist is high up in the tree Tarzan panics and starts to scream, so that his mother, Tepel, comes rushing up. Tepel herself never undertakes such foolhardy adventures, and she becomes extremely aggressive. When Puist has climbed down again and Tepel has Tarzan safely back, Tepel turns on the much larger and more dominant female to fight her. Yeroen rushes up to them, throws his arms around Puist's middle, and flings her several meters away.

This particular intervention was remarkable because on other occasions Yeroen had always intervened in Puist's favor. This time, however, he agreed, so to speak, with the protest of the mother and waived his usual preference.

Example 3

Before launching into the adoption experiment with Gorilla and Roosje we decide to show Roosje to the colony from behind one of the windows of our observation post. Roosje has been absent for six weeks. The whole colony bursts out hooting and collects below the window. The fiercest response comes from Yeroen, who normally never reacts to anything we do. He jumps up and down excitedly and throws sand and sticks at us. For three weeks he continues to be aggressive toward Monika, the keeper, whether she has Roosje with her or not. Yeroen is the only member of the group to behave in this way. Once we hand the baby over to Gorilla, his attitude becomes friendly again. We conclude

Luit (right) *was sitting by an attractive female* (left), *but now that Nikkie has driven him away and taken his place he is examining his fingernails with studied interest.*

that Yeroen was opposed to the fact that we humans had a chimpanzee infant in our keeping.

Example 4

Months later, when we introduce Gorilla and her foster child into the colony, we first of all let her walk past all the cages, in the morning before the chimpanzees are let out of their sleeping quarters, to gauge the reactions of the individual group members. Not only Krom, Roosje's natural mother, but also Nikkie reacts aggressively. We solve the problem of Krom by letting Mama in with her; Mama's presence immediately has a calming effect. Nikkie is more difficult. First of all, we release the whole group, except Nikkie. This introduction goes without a hitch. When Nikkie is let out a little later Yeroen and Luit join forces, with their arms around each other's shoulders, and form a barrier between Nikkie and Gorilla. Nikkie, who is the alpha male by this time, is chased away by a temporary coalition between the two older males. Afterward Nikkie runs screaming to Mama, then embraces Gorilla and kisses Roosje. That this male may have posed a lethal threat to the infant is quite possible given what we know now, twenty years later, about infanticide in the species: adult males occasionally kill newborns.[10]

These examples illustrate the great importance that males attach to the safety of infants. This protective attitude is more marked in the two senior males than in the two juniors. Once again this could be because older males have more offspring. They do not know who their children are, but their protective behavior increases the chances of survival of all possible offspring.

One problem remains. If I had to say which of the males is the most ambitious and jealous, I would say Yeroen. And with respect to protectiveness I would also say, after wavering between Yeroen and Luit, Yeroen. These characteristics are assumed to be associated with successful reproduction. It is, therefore, all the more remarkable that Yeroen wins on both counts when we know that despite frequent copulation (often with ejaculation) his physical deficiency does not allow him to fertilize any female. All his efforts are in vain. At first sight his case appears to disprove the theory.

In fact it does not do so, because Yeroen does not know the ultimate goal of fatherhood. He does not know that males can reproduce, because animals are not aware of the link between sex and procreation. They mate only for pleasure and are ambitious, jealous, and protective without knowing that this can benefit their offspring. Even though the function of their behavior, and the reason for its evolution, is to aid their offspring, they themselves only recognize certain *sub*goals: a high rank, more mating than other group members, and a safe environment for all the children in the group. They unconsciously serve the *main* goal of all living creatures. The fact that even an impotent male such as Yeroen does this illustrates the blindness of the urge to reproduce.

Sexual Bargaining

The oldest children in our colony were fathered in Yeroen's heyday, which means that he cannot have had a complete sexual monopoly. Lower-ranking group members always find ways, although this often means being secretive. Mariëtte van der Weel studied the openness with which the group members mated by recording which of the other males could see the act. She found that Nikkie and Dandy were concerned about their mating sessions being visible to others, whereas Yeroen and Luit were not unduly worried. We would have expected this of Luit, because he was the alpha male at the time, but not of Yeroen. I have never once seen Yeroen make a "date" with a female: he either mates openly or not at all. Perhaps this has something to do with the females. A conspiracy is needed to set up a mating session removed from the others. Perhaps the females, who accept Yeroen's direct invitations, are not prepared to walk a long way for sexual contact that, in Yeroen's case, is unsatisfactory.

The female is free to choose whether or not to have sex. Males sometimes put aggressive pressure on them, and in Arnhem we have in one case seen forced copulation (see note 7, p. 219), but normally, if the female does not want to mate, then that is the end of the matter. Persistent males run the risk of being chased by the female they approached and some of the other females too. Puist, who normally sides with the males in a conflict, always supports the estrus female in such cases. Hence the link between dominance and sexual rights, which exists among the males, is only half the story. Another important factor is the individual preference of the female, and this does not always tally with the rank of the male. Consequently it is the females who largely engineer the evasion of the rules that exist among males.

That everyone knows these social rules is clear not only from the furtiveness of certain contacts but also from the phenomenon of *telling tales*. Two examples serve to illustrate this. In the first, Dandy sees Luit paying court to Spin while the leader, Yeroen, is sitting a long way off and cannot see what is going on. Barking excitedly Dandy runs to Yeroen and attracts his attention. He then leads Yeroen to the spot where the two are in the middle of mating.

The second example dates from the period of Luit's leadership. While Luit has his back turned Yeroen and Nikkie both seize the opportunity to invite Gorilla to have sexual intercourse. She ignores Yeroen and presents to Nikkie. At once Yeroen begins to hoot at Luit, who turns round. Nikkie remains rooted to the spot and then wanders away as nonchalantly as possible.

Besides the secretive ways employed by the lower-ranking males to

Franje (left) *quietly awaits the outcome of the "bargaining" among the three males in the background.*

"enjoy a bit on the side," there are also occasions when they can mate openly by taking advantage of rivalry between dominant parties, or through "transaction" and "bargaining."

During the first period of Nikkie's leadership, the summer of 1978, it was Yeroen who mated most frequently. He kept both Luit and Nikkie away from estrus females by playing off his two rivals against each other. As soon as Nikkie approached the female or tried to intimidate Yeroen, the former leader appealed, screaming, to Luit to help him against Nikkie. Luit was only too willing to stop any rival. Conversely, when Luit ventured to approach the female Yeroen appealed success-fully to Nikkie. The jealousy between Nikkie and Luit was a powerful instrument in Yeroen's hand. (This situation confirms the importance of my earlier distinction between rank, or formal dominance, on the one hand and power, or social influence, on the other. Formally Yeroen dominated neither of the other two males, but his influence was clear enough, at least as far as sex was concerned.)

On September 5 everything was suddenly different. We saw Nikkie and Luit mating regularly and openly while Yeroen lay some distance away and did not interfere. The suddenness with which this turnabout took place was surprising. It seems likely that it was due to a "treaty" of nonintervention between Luit and Nikkie, by which they each "undertook" to give up supporting Yeroen against the other. Chimpanzees tend toward social reciprocity; hence Nikkie and Luit were each prepared to stop their interventions in Yeroen's favor in return for the other's neutrality. The result of such a wordless process is indistinguishable from that of a transaction.

The "treaty" against Yeroen was a return to the former open coalition between Luit and Nikkie. If no female was in estrus, there were no problems; Luit was extremely subservient and kept out of the way of the ruling Nikkie–Yeroen coalition. But in periods of sexual rivalry Luit underwent a startling metamorphosis. He walked around displaying self-assuredly and "greeted" Nikkie far less often. Sometimes the two of them even bluffed Yeroen away from a sexually attractive female. This might lead to conflicts that closely resembled those in the past: Luit and Nikkie united against Yeroen and his female supporters. In this way the old structure remained visible within the new one.

This all meant that our alpha male was a party to two coalitions, each with a different function. *Nikkie* used Yeroen's support to dominate Luit, and he used Luit's support, or at least neutrality, to prevent Yeroen from having sexual intercourse. *Yeroen* had regained a great deal of his former

Nikkie watches Luit as he approaches a female.

prestige, which he had lost to Luit, by helping Nikkie to power. *Luit,* finally, had strengthened Nikkie's position within the ruling coalition by withdrawing his support from Yeroen. Conversely, Nikkie's protection of Yeroen against Luit was minimal, particularly in periods of sexual rivalry. Hence Luit's influence in the group rose and fell with the swellings of the females.

Here we have a perfect example of a system based on the balance of power: the superiority of one party over another depends on the support of a third, so that each party affects the position of the others. Nikkie occupied the key position, which meant that the other two males both contributed to his power and shared in it. The power was disproportionately divided, but it was nevertheless not all in the hands of one individual. Indeed how could it be otherwise among animals with such a marked tendency to coalition formation? To quote Martin Wight, writing about international politics: "The alternatives to the balance of power are either universal anarchy or universal dominion." I cannot imagine either of these two alternatives in chimpanzee politics.

By 1980 Nikkie was mating about twice as often as the other two males put together. He broke up the others' mating sessions by displaying at the "wrongdoer," often with the third party at his side. But his own advances to sexually attractive females were not always left unchallenged. The other two males sometimes approached each other, hooting, so that there was always the threat of an intervention.

Sexual tolerance is influenced not only by the balance of power but also by calming efforts and grooming. Here is a typical scene: Franje is in estrus. The three males are sitting 10 meters away. Luit strolls up to Franje to inspect and smell her swelling. With his hair on end Nikkie goes and sits next to Luit, threateningly. Luit leaves the female. He grooms Nikkie, and after a while he invites Franje to mate. Franje hesitates because Nikkie has his hair on end again. Luit turns around with a broad grin on his face and holds out his hand to Nikkie. Then he returns to Nikkie to groom him again. When Luit invites Franje a second time the two of them are allowed to copulate undisturbed.

This is not an exceptional incident, but a very abbreviated description of what usually happens. When Rob Hendriks timed grooming sessions among the males with a stopwatch he found that they lasted nine times as long in periods when there was an estrus female in the group. What is the function of all this grooming? Perhaps its calming effect soothes away resistance in the grooming partner, so that he ultimately tolerates his rival having sexual contact. This would explain why the males always look hesitantly at their grooming partner as they approach the female, and also why they sometimes hold out a hand to him. This is a begging gesture. What, for example, could Luit have been doing at such a moment other than begging Nikkie to approve his sexual contact with the female?[11]

If grooming and other reassuring contacts are really maneuvers to prevent aggressive interventions, then we only have to go a little step further to call this the "price" and to talk of *sexual bargaining.* Even the alpha male has to pay this price. On occasion Nikkie has been seen to hold out his hand to the other two males. If they reacted to this gesture by displaying slightly or hooting, Nikkie returned to them and continued to groom, thus raising the price.

On one occasion Nikkie was so engrossed in grooming Yeroen that he did not notice Luit's silent departure. When, some time later, he glanced at the spot where Luit had been sitting, he screeched and looked round in all directions: the estrus female had also disappeared. Shocked, Nikkie and Yeroen embraced each other. They had obviously reached the same conclusion, because they both rushed wildly, their hair on end, across the enclosure and only calmed down when they found Luit quietly drinking from the moat. Had their fears been unfounded? They will never know, but I know that they had simply arrived too late.

SOCIAL MECHANISMS

IN THIS FINAL CHAPTER I WILL DISCUSS A FEW GENERAL PRINCIPLES relating to life in the group and the mental capacities of the group members. On the one hand I will be discussing capacities such as strategic intelligence, which we suspect but cannot in fact prove. On the other hand there are capacities that we tend to take for granted, but that are not necessarily present in animals. The most basic of these is the ability to recognize each other individually. Although animals who do not have this ability can form a hierarchy, they are forced to reassert their position within this hierarchy at each meeting. Individual recognition removes this uncertainty and ensures a well-established structure in which everyone knows his place. If the number of individuals becomes unnaturally large, the system collapses. For example, some monkey groups in Japan have expanded to as many as a thousand individuals because they are being fed by tourists. The normal number in a group would be around a hundred. When groups become as large as this a strikingly high number of unclear and unstable dominance relationships occur. Obviously the vast number of individuals exceeds the memory capacity of the Japanese macaque, so that many group members remain strangers.

Just as individual recognition is a prerequisite of a stable hierarchy, so *triadic awareness* is a prerequisite of a hierarchy based on coalitions. The term *triadic awareness* refers to the capacity to perceive social relationships between others so as to form varied triangular relationships. For example, Luit knows that Yeroen and Nikkie are allies, so he will not provoke conflicts with Yeroen when Nikkie is nearby, but he is much less reluctant to do so when he meets Yeroen alone. What is special about this kind of knowledge is that an individual is not only aware of his or her own relationships with everyone in the group, but also monitors and evaluates relationships that exist in the social environment so as to gain an understanding of how the self relates to *combinations* of other individuals. Elementary forms of three-dimensional group life are found in many birds and mammals, but primates are undoubtedly supreme in this respect. Mediation with a view to reconciliation, sepa-

The inflated display is usually a short, self-assured charge. The facial expression, with bulging lips, that goes with it is difficult to photograph. I succeeded in taking this photo by calling out Nikkie's name immediately after he sat down following a charge.

rating interventions, telling tales, and coalitions would all be inconceivable without triadic awareness.[12]

This mental ability is naturally reflected in all social domains. The following example stems from a nonpolitical context. One day Franje takes Jimmie's child. The child utters a brief scream, whereupon Jimmie jumps up and hurries over toward Franje. About 15 meters away from Franje she sits down threateningly with her hair on end. Franje looks at Jimmie but seems to be frightened of coming any nearer. At that point Amber intervenes and solves the problem. She takes the child from Franje's back and carries him back to his mother. As Amber hands over the child to Jimmie, Franje "greets" Jimmie submissively, but from a great distance.

To intervene successfully Amber not only had to recognize Jimmie, Franje, and the child as individuals, but to understand the cause of the conflict she also had to know to which of the two females the child belonged. If this sounds simple it is because triadic awareness is second nature to human beings, and we find it hard to imagine a society without it.

Dependent Rank

The affection between my two tame jackdaws is not mutual. The older bird, Rafja, was brought up by me without the company of other members of her species. Socially and sexually she is more interested in me than in the younger bird, Johan, who is courting her. Because Rafja rejects Johan this gives rise to conflicts during the mating season. Johan is much bigger and stronger than Rafja. When I hear her screaming I rush to the aviary to save her from his claws. When I arrive, Rafja flies to my shoulder and screeches aggressively at Johan. Sometimes she even attacks him from her safe position and pecks at him. Instead of fighting back, Johan flees from her. This means that their dominance relationship is reversed by my intervention. Konrad Lorenz described a similar phenomenon in 1931. In his first lengthy publication he observed how the females in his tame jackdaw colony rose in rank by virtue of their pair bond, taking on the same status as their male partners.

It was not until 1958, however, that the phenomenon received a name, when it was discovered, completely independently, among Japanese macaques. Masao Kawai threw down some food for two young

Japanese macaque with infant.

monkeys living in a wild troop and recorded the outcome. He regarded the monkey who took the food as the dominant one. Kawai conducted a large number of such tests and found that some dominance relationships depended on the distance of the infants from their mothers. For example, monkey A dominated his peer B when their mothers were far off, but the converse was true when the mothers were nearby. These reversals occurred if B's mother dominated A's mother. The offspring of a high-ranking mother benefits from her physical presence. Kawai called the relationship between the two infants in the absence of their mothers the *basic rank,* and in the presence of their mothers the *dependent rank.*

If we apply the same terms to my pet jackdaws, Rafja is basically subordinate to Johan, but my presence elevates her dependent rank. This phenomenon will, of course, presuppose a nonneutral presence. Both in Rafja's case and in the case of young monkeys the rank depends on an individual who is prepared to protect them. Triadic awareness plays an important role here. From experience the conflicting parties know on whose side the third party stands, consequently the third party can influence the situation by a mere look or gesture of support.

Dependence on third parties plays such a prominent role in the chimpanzee hierarchy that the basic relationships are completely overshadowed. This is not only true for the complex balance of power in the male triangle. A small child, for example, may chase away a full-grown male. He is able to do so under the protection of his mother or "aunt." Like the children, these females are basically inferior to the males, but they, in turn, can rely on the support of other females and sometimes can appeal to dominant males for help.

In their wild habitat, where the females live far more scattered, it is extremely rare for a male to be driven away by a child or a female. The difference in power is much smaller in our colony because of the limited space they live in and the relatively large number of females.[13]

The Female Hierarchy

The basis of hierarchical positions is sex-related. Among males coalitions determine dominance. The male dominance over the females is largely determined by their physical superiority. Among females it is above all personality and age that seem to be the determining factors.

Conflicts between females are so rare and the outcome is so unpredictable that they cannot be used as a criterion for determining rank. This has also been observed in Gombe, where David Bygott concluded: "It may not be very meaningful to describe female-female antagonistic relationships in terms of dominance." In our colony the number of male-male conflicts is one every five hours, the number of male-female conflicts is one every thirteen hours, but the number of female-female conflicts is only one every hundred hours (this is the average frequency of both major and minor conflicts per combination of two individuals during summer).

"Greetings" between females occur twice as frequently as conflicts. Even this is very little, but of all the possible criteria only "greetings" constitute any indication of hierarchy. Looked at in this way the female hierarchy can be divided into four classes: Mama, the alpha female, is in a class on her own; then Puist and Gorilla; then the first three mothers of the colony, Jimmie, Franje, and Tepel; and under them Spin, Krom, and Amber. Between the different classes the relationships are much clearer than within the classes. The great difference between the male and female hierarchies is that the female hierarchy has been stable for years.

Intensive observation of a small captive group of macaques will reveal the female hierarchy within a few days. In the case of our chimpanzees we have to allow many months to arrive at the same conclusions. The vagueness of the female hierarchy is largely due to the lack of assertiveness. Whereas female macaques and baboons, along with male chimpanzees, regularly prove their dominance, female chimpanzees do not seem to have the same need. The apparent lack of female ambitions is probably connected with the fact that chimpanzees tend not to live in such a close community and usually search for food on their own. There is much less competition for food than in the tight-knit groups of macaques and baboons. This means that an important reason for striving to be higher in rank is absent in free-living chimpanzees, and the females in our colony reflect this pattern.

The female hierarchy in our chimpanzee group seems to be based on respect from below rather than intimidation and a show of strength from above. Females seldom display, and 54 percent of the "greetings" among them are spontaneous, against as little as 13 percent among males. As among females of other great ape species, acceptance of dominance is probably more important than proving dominance. For example, when orangutan females are placed in a cage together for the first time they immediately, without the least hesitation, establish a stable dominance pattern, without fighting or threatening in any way. This dominance is not determined by their physical size; maybe the personality of certain individuals inspires respect. In the field, seniority and residency seem to be critical factors. Perhaps female apes quickly evaluate a number of such factors and submit readily to females who seem to have the upper hand. It may be a mistake, however, to conclude from the quick establishment and subsequent stability of dominance that female apes are totally indifferent to which rank they occupy. Rare observations of fierce competition warn against such a conclusion.[14]

Our understanding of ape hierarchies is further complicated by the fact that there is a third type of dominance that exists alongside formal dominance and power. For example, when the alpha male places a car tire on one of the drums in the indoor hall with the intention of lying down on it, one of the females may push him away and sit down herself. Females also remove objects, sometimes even food, from the hands of the males without meeting with any resistance. One of my students, Ronald Noë, compared three types of interaction between the senior males and high-ranking females. According to the formal criterion of who "greets" whom, males dominated 100 percent of the time; and according to the criterion of who wins aggressive incidents, 80 percent of the time. As far as taking away objects and places to sit were concerned, however, females dominated 81 percent of the time. Since females lack the physical attributes to claim resources by force, their precedence must depend on male tolerance. But then the question arises why males would allow females to act this way. Could it be that females bring weapons other than physical strength to the battlefield? They have things to offer that cannot be taken by force, such as sexual and political favors, and their silent diplomacy, which helps to calm tempers. This provides the females with a good deal of leverage: if being popular among the females is critical for the stability of a male's leadership, he had better be lenient and accommodating toward them.

Visitors to the colony always like to know who is the boss, and I take pleasure in confusing them by saying, "Nikkie is the highest-ranking ape, but he is completely dependent on Yeroen. Luit is individually the most powerful. But when it comes to who can push others aside, then Mama is the boss." The anthropologist Marshall Sahlins thought that such exceptions to the law of the strongest were unique to the human race: "Quite the opposite from subhuman primates, a man must be generous to be respected." If Sahlins had baboons or macaques in mind, he most certainly was right, but in the case of our nearest relatives, the anthropoid apes, things are considerably more complex and more human-like. Recently, Toshisada Nishida described the case of an alpha chimpanzee in the Mahale Mountains who maintained his rank for an extraordinarily long time (over a decade) through an elaborate system of "bribery." He distributed meat selectively to those individuals whose support he could use against potential challengers.

Strategic Intelligence

Ever since Thucydides wrote about the Peloponnesian War, more than two millennia ago, it has been known that nations tend to seek

Like Nikkie before him, Dandy is going through a phase of bullying the females.
He will only stop being aggressive when they begin to respect him through
pant-grunting. On the left, Dandy; on the right, Mama.

allies against nations perceived as a common threat. Mutual fear as the
basis of alliance formation makes nations weigh in on the lighter side
of the balance. The result is a power equilibrium in which all nations
hold influential positions. The same principle applies to social psychology and is known as the formation of "minimal winning coalitions." If
the weakest of three players in an experimental game has a chance of
scoring points if he joins the strongest or the middle party, he will prefer
to ally himself with the latter. After his dethronement Yeroen was faced
with a similar choice; on the one hand a coalition with the more powerful party, Luit, and on the other a coalition with the weaker, Nikkie.
Under Luit's dominance Yeroen's influence was limited because Luit
did not need his support. At most he needed his neutrality. By choosing to support Nikkie, however, Yeroen made himself indispensable to
Nikkie's leadership, and consequently his influence in the group grew
again.

*Luit, nearly in a trance, with his eyes closed, dances a rhythmic stamping dance
and crows during the climax of a ventilating display.*

If Yeroen's strategy has much in common with that of countries and
individual humans, we must ask ourselves whether the background to
his behavior may not be the same. Among humans, strategy is based
on rationality. This should not be confused with consciousness: we can
unconsciously arrive at rational solutions, and we also sometimes take
steps that we know are irrational. A rational choice is based on an *estimate of the consequences.* Thus, the question is whether Yeroen looked
into the future before deciding to join forces with Nikkie.

The first thing we need to determine is whether chimpanzees actively
strive for a higher rank. Does the urge to achieve it provoke what we

term goal-directed behavior? The chief characteristic of such behavior is that it becomes superfluous after the goal has been achieved. Behavior X is said to serve goal Y, if X stops when Y is reached. The simplest example of such functioning is the thermostat. This causes the oven to heat up (behavior X) until a certain temperature is reached (objective Y), then it cuts out. Jane Goodall's observation about Goblin, a Gombe Stream chimpanzee, provides a clear parallel: "In adolescent male chimpanzee style, he had bullied and blustered at the adult females until they began to defer to him." The word "until" is crucial here. Why did Goblin no longer behave in this way once the females had deferred to him? We have a similar case with Nikkie. The figures show that his hostility toward the females started to decrease in the autumn of 1976, the time when they began to "greet" him regularly. We can therefore deduce that the behavior of an adolescent male mellows with female recognition of his dominance. The simplest explanation for this is that his provocative actions are a way of forcing others to respect him, and that they become superfluous as soon as this goal is achieved.

If this sounds perfectly natural, we must not forget that ambition among animals has long been a subject of controversy. In 1936 Maslow postulated a *dominance drive,* but most ethologists have carefully avoided the term. From my own study of both macaques and chimpanzees I have no hesitation whatsoever on this point. The animals I have observed clearly strove to attain a higher status. To quote Jane Goodall once again: "Quite clearly, many of the male chimpanzees expend a lot of energy and run risks of serious injury in pursuit of high status." Evidently the lust for power is not confined to animals in zoos.

Java macaques can threaten each other in two ways. When I was studying them I found that there was a difference in function between these two different forms. Young monkeys use a noisy form of threatening against individuals whom they will later supersede in rank, that is, individuals whom their mothers dominate. The other, quiet form of threatening is used against individuals who have already shown themselves to be subordinate. The first form of threatening is used to climb the social ladder, while the second form merely underlines existing positions.

A similar difference is seen among chimpanzees in their displays. Once again the difference is in the amount of noise produced: one is deafening, the other is silent. The first form of display begins with the swaying of the upper body and a gradually rising hoot. After this the male rushes past his rival, thumps the ground and finishes with a loud crow. Because of the deep, rhythmic inhalation and exhalation that accompanies this hooting, this is called *ventilating display.* In all dominance processes this kind of display was particularly marked in the challenging party. Once the period of instability was over and the rival had submitted, the challenger switched to the other form of display. In this form the lips are pressed firmly together and the male holds his breath: the chest is puffed up and the lips bulge under the pressure. This is called *inflated display.* Whereas ventilating display is challenging and serves to make aspirations known, inflated display is self-confident and assertive.

In the same way that serious fights decrease in number after a dominance reversal, so do ventilating displays. This supports the claim that these processes are goal-directed, because otherwise why should the deference of one party be answered by a change in bluff style by the other party, and why should there be a decrease in the number of confrontations? What is more, the fact that males often get into conflict without any obvious reason, whereas females and children tend not to, suggests that rivalry about something as "intangible" as status is their prime motive. It is my opinion, therefore, that chimpanzees, and certainly male chimpanzees, actively strive for a higher status.

The hankering for power itself is almost certainly inborn. The question now is, how do chimpanzees achieve their ambitions? This too may be hereditary. Some people are said to have "political instinct," and there is no reason why we should not say the same of chimpanzees. I doubt, however, whether this "instinct" is responsible for all the details of their strategies.

Experience is needed to use innate social tendencies as a means to an end in the same way that a young bird born with wings to fly needs months of practice before it has mastered the art. In the case of political strategies, experience can play a role in two ways: directly, during the social processes themselves, or through the projection of old experiences onto the future. The first possibility means that an ape such as Yeroen has noticed that by supporting Nikkie he reaps certain benefits. This could be conditioning: a particular behavior is reinforced by its positive effects. This cannot be the whole answer, however, because the consequences of Yeroen's strategy were initially *negative.* By resisting

Luit and seeking contact with Nikkie, Yeroen found himself in conflicts that he usually lost. For Yeroen to have continued his policy he must have been certain, inwardly, that he was on the right track, because it was not until months later that his choice began to show obvious advantages. Yeroen may have been stimulated either by subtle effects (for example, signs of uncertainty in Luit) or by a *prediction* as to the eventual outcome of the process. The ability to make predictions is something we hardly dare to attribute to animals. While it cannot be proven in Yeroen's case, there is evidence that chimpanzees possess the requisite mental faculties.

Dalbir Bindra defined planning as the "identifying of a 'route' of subgoals that links the subject's present position to the remote goal." This requires an ability to look into the future. Bindra observes that: "A chimpanzee is probably capable of planning that spans a longer period of time, as well as capable of separating the planning from the execution of the plan, at least to some extent." I will give two examples of such future-oriented behavior.

It is November and the days are becoming colder. On this particular morning Franje collects all the straw from her cage (subgoal) and takes it with her under her arm so that she can make a nice warm nest for herself outside (goal). Franje does not do this in reaction to the cold, but before she actually feels how cold it is outside.

The second example is "saying goodbye." In contrast to greeting, which is a reaction to meeting someone familiar, saying goodbye rests on anticipation of separation. I received the first hint that it is not impossible for apes to foresee a parting at a conference in Germany, where Allen Gardner gave a lecture about language experiments with young chimpanzees. The apes were taught to use hand signs, and they communicated in this way not only with humans but also among themselves. Gardner told us that they used the "bye bye" signal before they separated. Later I received a strong indication from our own colony: Gorilla has to bottle-feed Roosje every afternoon, and the keeper calls her indoors at the appointed time, while the rest of the colony stay outside. Before she goes inside with her foster child, however, she walks over to Yeroen and Mama and touches them briefly or kisses them. Sometimes this involves her making quite a detour. The only explanation for this behavior seems to be that she is saying goodbye, because she can see ahead to the separation.

It is important to realize that such future-oriented behavior is based on experience. It is quite different from, for example, the precautions taken by squirrels in the autumn in gathering food for the winter, because even squirrels who have not yet experienced a winter do this. What is more, chimpanzees are able not only to look and think ahead but they can also look several steps (subgoals) ahead. Proof of this was supplied by an ingenious experiment by Jürgen Döhl.

Julia, a female chimpanzee, was presented with a box containing two keys, from which she had to choose one. With this key she opened another box containing the key to yet another box, and so on until she would eventually arrive at the last box. If Julia had chosen the right key at the beginning this last box would contain a tasty tidbit. If, however, she had chosen the wrong key she would embark on a route of boxes and keys which would lead her to an empty box. Julia had been taught which keys fitted which boxes, and because the boxes were transparent, she was able to see which keys lay in which boxes. She had to suppress her impulse to take pot luck in choosing the first key because if she chose the wrong one she was not allowed to rectify her mistake, and in the next trial there would be a completely different arrangement of keys and boxes. Hence Julia was forced to connect the ultimate goal, the food box, with the initial choice of key, and it made her think before she chose. In experiments involving ten boxes all mixed up (two series of five) Julia was able to do just that. She was in fact able to look several steps ahead to achieve her goal.

Later, Döhl added that chimpanzees in their natural habitat are never faced with problems as difficult as this one. As so often occurs in experimental psychology, this comment reflects a lack of appreciation of the complexities of real life for which the mental capacities measured in the laboratory are an adaptation. Biologists are extremely skeptical of any possibility of superfluous capacities in animals: why would nature build and maintain an energetically expensive instrument, such as a large brain, if not for increased survival and reproduction? Obviously, chimpanzees are normally never faced with the exact same technical problem as presented to Julia, but we cannot exclude that the sort of strategic intelligence that she demonstrated is crucial in the *social* domain. It is exactly this ability to consider a remote goal and weigh the consequences of a choice that could explain why a male, such as Yeroen, formed the alliance that ultimately offered the best prospects.

It often happens that humans discover the goals of their behavior only in retrospect. During adolescence, for example, we stand up against our parents, provoking and challenging them. Later we may explain this behavior by saying, "I wanted my independence," but remember that we did not start the generation conflict with this motive explicitly in mind. It was an unnamed, unconscious motive. Similarly, we may want to influence others, and develop tactics to do so, without ever realizing for ourselves that this is our goal. We may even avoid thinking and talking about it. A Dutch social psychologist, Mauk Mulder, has conducted many experiments that have shown that men get satisfaction from wielding power and that they strive to increase their influence over others. At the same time, however, he points out that there is a taboo surrounding the word "power": "When we talk of power, we are talking about someone else . . ." When referring to ourselves we prefer to speak of "carrying responsibility," "being in a position of authority," or "helping others by taking decisions out of their hands."

A cat stalking a bird has to "calculate" her final jump. She needs a lot of experience to predict accurately the bird's reaction; there is an enormous difference between young and adult cats in this respect. We do not generally regard a cat's calculations as a conscious process, and I think that our own strategies largely depend on very similar intuitive calculations. They are based on experience and demand much intelligence but do not necessarily penetrate into our conscious mind. In the same way, chimpanzees may follow a rational course of action, making use of their intelligence and experience, without consciously planning their strategy.

Humans are talking primates, but in fact their behavior is not very different from that of chimpanzees. People engage in verbal fights, provocative or impressive word displays, protesting interruptions, conciliatory remarks, and many other patterns of verbal activity that chimpanzees perform without an accompanying text. When humans resort to actions instead of words the resemblance is even greater. Chimpanzees scream and shout, bang doors, throw objects, call for help, and afterward they may make up by a friendly touch or embrace. We humans also display all these patterns, usually without taking a conscious decision to do so, and maybe our motives are not so very different from those of chimpanzees.

Differences between the Sexes

Some ethologists hesitate to speak of sympathy between animals, but outsiders who are told of the far-reaching collaboration between primates have no such inhibitions. The experimental psychologist Nick Humphrey writes: "The selfishness of social animals is typically tempered by what, for want of a better term, I would call *sympathy*. By sympathy I mean a tendency on the part of one social partner to identify himself with the other and so to make the other's goals to some extent his own." It is only possible to prove this intuitive interpretation if we have an independent measure for sympathy. Let us say that sympathy is related to intimacy and familiarity, which can be measured by the amount of time two individuals spend together.

In the case of Congo, the famous painting chimpanzee, there was certainly a link between intimacy and the degree of support given. Desmond Morris reports that Congo developed a set of friendship priorities with four people, and that these priorities corresponded with the amount of contact he had with each of them: "The four of us then experimented by pretending to attack one another. In each case Congo defended the one he knew best. The intriguing thing about his loyalty rating was that it made no difference to him which of two people was the attacker and which the attacked." Congo's interventions followed his sympathies.

Do chimpanzees do the same thing with respect to one another? There is sufficient material available to answer this question as far as the Arnhem colony is concerned. First of all, over the years nearly five thousand interventions in conflicts have been registered and so we know which individuals support whom. We also know which partners each individual prefers to groom, play, walk, and sit with. The following example shows how we can compare these two factors. A quarrel between Tepel and Amber is interrupted by Puist who attacks Amber. According to our records, Puist has far more contact with Tepel than with Amber. This means that she intervened, like Congo, in favor of the more familiar party. With regard to the total number of interventions by all individuals in the group, this happens in 65 percent of the cases. If familiarity were not a determining factor, the percentage would be around the 50 percent mark. This confirms the importance of personal preferences.

Calculations and predictions are practiced young. Moniek knows from experience that dead branches break and demonstrates her knowledge by lifting up her left foot, ready to catch the end of the branch.

At first sight this result is trivial; who would have expected anything else? The surprise comes when we look at each individual separately. Of the twenty-three individuals, twenty-one score over the 50 percent limit, and only two score below. The exceptions are Yeroen and Luit. The vast difference between them and the rest cannot be merely chance; the number of interventions recorded is large enough to give a reliable result. This means that the oldest males do not allow personal preferences to influence their interventions. My interpretation is that Yeroen and Luit are extremely aware of the political effect of what they do. Their interventions are in accordance with a policy directed at increasing power. The flexibility with which they make and break coalitions gives one the impression of policy reversals, rational decisions, and opportunism. There is no room in this policy for sympathy and antipathy.

This interpretation is supported by the finding that Yeroen and Luit

showed their lowest percentages during periods of unstable leadership, whereas they scored above the 50 percent mark during stable periods. And also Nikkie went through a period when his interventions did not correspond with his friendships: this was during his rise in rank over the females and Yeroen. The conclusion is that collaboration is not always based on sympathy, especially not during status competition among adult males. Females and children, on the other hand, usually show sympathy-biased interventions: as a group they score about 75 percent. They largely intervene in conflicts to help relatives and good friends, that is, they react to events in the group rather than use intervention as a means to achieve dominance. The contrast between the sexes cannot be denied. Stated in the simplest terms, the one is protective and personally committed, the other is strategic and status oriented. The picture looks familiar? Am I allowing myself to be governed by my prejudices, or is this once again a striking similarity between chimpanzees and humans?

The photographs on pp. 190–94 show tool making, tool use, and cooperation in order to reach fresh leaves on trees that are protected by electric fencing. Opposite, high up in one of the dead oak trees, Nikkie is trying to break off branches; *above,* he is pushing apart the two ends of a forked branch.

Above, *one of the branches falls out of the tree and breaks in two. Luit holds one part, which is too short for the purpose, while Nikkie, still up in the tree, breaks off and carries down another branch.* Opposite: *Luit takes the long branch from Nikkie (top) and carries it to the protected trees while the others follow him, hooting excitedly. Luit props the branch against the tree (bottom left), with the forked end on the ground;* bottom right, *Nikkie now places the fork upward, giving more support.*

While Nikkie holds the branch tightly, Luit climbs into the tree.

Sex differences used to be a controversial topic insofar as humans were concerned. The debate revolved around innate versus environmental origins, with many feminists and social scientists emphasizing the latter and minimizing the former. This dichotomy made most biologists uncomfortable: we believe that everything humans do is determined by a combination of influences. Presently, public opinion is coming around to this insight. Men and women differ genetically, anatomically, hormonally, neurologically, and behaviorally; it simply does not make sense to isolate the last difference from the other four. Having said this, it is also clear that behavioral differences do not need to follow from genetic programs that "dictate" the behavior of each sex. Such instinctivistic explanations cannot even begin to deal with the personality differences observed, nor with the obvious influence of the social environment.

In the course of time, I have changed my opinion about sex differences in chimpanzees from explaining them as inborn behavioral tendencies to viewing them as reflecting different social objectives in males and females. If the sexes are trying to get something different out of life we obviously expect different behavior: the road to goal X requires another behavioral strategy than the road to goal Y. These strategies need not be genetically specified; they may develop through experience and learning.

Female chimpanzees tend to avoid competition. The distribution of their food in the wild requires them to disperse over the forest so as to get enough to eat. They also have an interest in creating a safe environment to raise their offspring, which would explain their concilatory role in a situation, such as the Arnhem Zoo, in which they are forced to live together with adult males. It also explains their support for older, established leaders instead of young upstarts, such as Nikkie. Social stability is to their advantage. For males, on the other hand, stability is only a good thing if they are at the top. Males evolved a tendency to establish a close-knit, hierarchically organized group within which to seek dominance over others. Males depend on one another in the wild; cooperation is a matter of life and death during territorial encounters with neighbors. The male's reproduction is served by maximizing the number of available females, which depends on the size of the community's territory, and the number of offspring sired, which depends on their social rank within the system.

What is known about sex differences in human coalition behavior? Two social psychologists, John Bond and Edgar Vinacke, organized groups of three people—two men and one woman or two women and one man—to take part in a competitive game where the formation of coalitions increased the chances of winning. After 360 such games they concluded that men take more initiatives with respect to forming alliances, especially if there is something to be gained by doing so, whereas women find the atmosphere in which the game is played more important. Women support weakly positioned players and join forces against male competitiveness. The female strategy yields similar results, but is so totally different that the psychologists call their coalitions *accommodative* as against the *exploitative* coalitions of the men. A great many such studies have been conducted, and they all point in the same direction: men are out to win and concerned with strategic considerations, whereas women are more interested in interpersonal contacts and predominantly form coalitions with the people they like.

Exploitative coalitions and the opportunism that goes with them are most clearly seen in politics. Anthropological and politicological studies show that women are interested in local rather than distant political affairs and that men prefer to tackle "big" politics and are attracted to the center of power. Because this difference between the sexes is universal, J. Dearden, who compiled literature on the subject, places the emphasis on biological rather than socio-cultural factors. Differences between the sexes are always of a statistical nature, however. We all know exceptions to the rule in human politics, and also in the chimpanzee world there are definitely exceptions. Males who see that they have a chance of success by collaborating with very familiar partners will certainly do so. The coalition between the brothers Faben and Figan in Gombe is an example of this. Conversely, females are not always as completely uninterested in power as has been suggested. Mama did not relinquish her former top position without resistance, and in the power takeovers between the males, particularly the first one, the females played an active role.

The fact that a large number of females constantly live together in our colony gives them greater political influence than in chimpanzee communities in the wild. If the environment is so important in determining the role of the sexes, just how important is the biological element? Hugo den Hartog, a sociologist, once answered my question with the following "minimum proposal": the top is attainable for high-ranking individuals, and they are therefore motivated to carry out aggressive strategies and intimidations; the lower echelons, on the other hand, do not have this stimulus and react to their suppression with stable, loyal, and friendly cooperation. Male chimpanzees constitute the first category by virtue of brute strength, and they act accordingly. In other words, the genetic influence is restricted to a difference in physical strength, which, because it creates a gap in the hierarchy, has enormous consequences as far as role differentiation is concerned.

This hypothesis is very attractive. The only shortcoming is that it does not explain the sex differences in human coalition behavior, which are so strikingly similar to those in chimpanzees (unless differences in size and strength between men and women play a more important role in our society than we are accustomed to think). Anyway, although behavior is never hereditary from A to Z, the genetic influence may be greater than has been suggested. In order to separate the two, an experiment with two groups is needed: a purely female and a purely male group. The females at the top of the hierarchy in the first group will be in the same situation as the one in which males develop exploitative strategies. Would such a situation make opportunists of females too? On the other side of the coin we would be able to see whether low-ranking males in the second group would adopt the same protective, loyal attitude as females in comparable positions. Without having conducted such an experiment I do not dare to predict what the outcome would be.

Sharing

A short while ago we threw a large amount of oak leaves out of our observation window. Seeing that Yeroen was approaching at full speed, bluffing as he came, none of the other apes dared to go near the leaves. Yeroen gathered up the whole pile, but ten minutes later each member of the group, from large to small, had a share of the booty. For the adult male, the amount that he himself possesses is not important. What matters is who does the distributing among the group. (However, this only applies to incidental, extra food. Main meals and hunger can cause chimpanzee males to quarrel violently, as the Holloman colony showed.) Females, on the other hand, tend to share mainly with their own children and best friends and do get into quarrels with other group members. Taking food by force is extremely rare in our colony; sharing is something apes learn young. An example: Oor has found a branch with leaves on it, and Fons is pulling at it, yelping, but gets nothing. Oor's best friend, Amber, approaches the quarrelling pair, takes the branch out of Oor's hands, breaks off a bit for Fons and a little bit for herself, then she returns the largest share to Oor.

Chimpanzees in their natural habitat are known to share meat after hunting. Adult males prey, sometimes in perfect cooperation, after which other members of the group come begging for their share. Concerted action and sharing of the yield are also common features in the Arnhem colony with the "hunt" for forbidden foliage. The males use long branches to climb up into the live trees, which are protected by electric fencing. At first branches that were lying around were used, but later branches were deliberately broken off the dead oak trees. It is an extremely hazardous undertaking to force two heavy, forked branches apart some 20 meters above the ground. The art is to keep a firm hold on the right side while breaking the branch, but not to let go of the side that is breaking off altogether. If a mistake is made, either the broken branch will come crashing down in pieces or the ape himself will fall. Luckily none of the apes has yet fallen, although they obviously have difficulty in predicting which side of the fork offers a safe hold. Sometimes they are frightened by the sound of cracking beneath them and quickly climb out of the tree with a yelp.

If everything goes according to plan, the male carries the branch down to the ground and sets it up as a "ladder," usually in close cooperation with the other males and sometimes the females. The ape in the tree breaks off far more than he needs, and this falls down among the waiting group. Sometimes the process of sharing is selective. Once, when Dandy held the branch steady so that Nikkie could climb into the tree, he later received half the leaves Nikkie had collected. This appeared to be a direct payment for services rendered.

It often seems as if the males invite one another to collaborate on some task in order to work off their joint tensions and frustrations. Luit, for example, used to climb into a dead oak tree to break off branches and initiate concerted actions, particularly when Nikkie had forbidden him to have any contact with Yeroen. "Discharge cooperation" of this kind may also occur opportunistically. For example, once Nikkie was displaying heavily at Yeroen and Luit, who were sitting together, and accidentally broke off a branch with one of his huge leaps. The rivalry was instantly forgotten. The three males embraced one another, then carried the broken branch to the live trees.

Holding a branch for another group member seems to be something people find more convincing as a proof of calculated assistance than the formation of a coalition. The same applies to sharing out leaves and meat. We see this as an act of generosity, more readily than the relinquishment of sexual privileges or the offer of protection. And yet both forms of giving are related. Chimpanzee males are surprisingly generous when it comes to material things; they even allow certain females to take objects from them. But they also demonstrate this characteristic in their social behavior (except, that is, with respect to rivals). Their control rests on giving. They give protection to anyone who is threatened and receive respect and support in return.

Also among humans the borderline between material and social generosity is scarcely distinguishable. Observations of human children by the psychologists Harvey Ginsburg and Shirley Miller have demonstrated that the most dominant children not only intervene in playground fights to protect losers but also are more willing to share with classmates. The investigators suggest that this behavior helps a child to command high status among peers. Similarly, we know from anthropological studies of pre-literate tribes that the chief exercises an economic role comparable to the control role: he gives and receives. He is rich but does not exploit his people, because he gives huge feasts and helps the needy. The gifts and goods he receives flow back into the community. A chief who tries to keep everything for himself puts his position in jeopardy. *Noblesse oblige,* or, as Sahlins said, "A man must be generous to be respected." This universal human system, the collection and redistribution of possessions by the chief, or his modern equivalent, the government, is the same as that used by chimpanzees; all we have to do is replace "possessions" by "support and other social favors."

It is perfectly reasonable to suppose that our forefathers had a centralized social organization long before material exchange began to play a role. The first system may well have served as a blueprint for the current one.

Reciprocity

The influence of the recent past is always overestimated. When we are asked to name the greatest human inventions we tend to think of the telephone, the electric light bulb, and the silicon chip rather than the wheel, the plough, and the taming of fire. Similarly the origins of modern society are sought in the advent of agriculture, trade, and industry, whereas in fact our social history is a thousand times older than these phenomena. It has been suggested that food sharing was a strong stimulus in furthering the evolution of our tendency to reciprocal relations. Would it not be more logical to assume that social reciprocity existed earlier and that tangible exchanges such as food sharing stem from this phenomenon?

There are indications of reciprocity in the nonmaterial behavior of chimpanzees. This is seen, for instance, in their coalitions (A supports B, and vice versa), nonintervention alliances (A remains neutral if B does the same), sexual bargaining (A tolerates B mating after B has groomed A) and reconciliation blackmail (A refuses to have contact with B unless B "greets" A). It is interesting that reciprocity occurs in both the negative and the positive sense. Nikkie's habit of individually punishing females who a short time before joined forces against him has already been described. In this way he repaid a negative action with another negative action. We regularly see this mechanism in operation before the group separates for the night. This is the time when differences are squared, no matter when these differences may have arisen. For example, one morning a conflict breaks out between Mama and Oor. Oor rushes to Nikkie and with wild gestures and exaggeratedly loud screams persuades him to attack her powerful opponent. Nikkie attacks Mama, and Oor wins. That evening, however, a good six hours later, we hear the sound of a scuffle in the sleeping quarters. The keeper later tells me that Mama has attacked Oor in no uncertain manner. Needless to say Nikkie was nowhere in the vicinity.

Negative behavior hardly enters into the theories about reciprocity that anthropologists and sociobiologists have developed. The titles of famous publications on the subject are sufficient proof. In 1902 Peter Kropotkin wrote *Mutual Aid: A Factor of Evolution,* in 1924 Marcel Mauss wrote *The Gift,* and in 1971 Robert Trivers wrote "The Evolution of Reciprocal Altruism." Despite the emphasis on positive exchanges there has been much theoretical progress. Some anthropologists deny the biological roots of human cooperation and solidarity, but the integration of cultural and genetic explanations cannot be resisted for much longer.

Another powerful school of thought is social psychology. Since the publication in 1959 of J. Thibaut and H. Kelley's *Social Psychology of Groups,* with its assertion that "every individual voluntarily enters and stays in any relationship only as long as it is adequately satisfactory in terms of rewards and costs," interactions between humans have been regarded as a kind of trading in advantageous and disadvantageous behavior. Here too reciprocity is an important theme, not only in its positive form but also in its negative form.

In short, scientists of all colors and schools of thought are fascinated by give-and-take arrangements; this must surely mean that the give-and-take mechanism is a very fundamental one. Whether what is involved is the returning of a favor or the seeking of revenge, the principle remains one of exchange; and, most importantly, this principle requires that social interactions be remembered. Much of the time the process may take place in the subconscious, but we all know from experience that things come bubbling up to the surface when the difference between costs and benefits becomes too great. It is then that we voice our feelings. By and large, however, reciprocity is something that takes place silently.

The principle of exchange makes it possible actively to teach someone something: good behavior is rewarded; bad behavior is punished. A development in the relationship between Mama and Nikkie demonstrates just how complex such influencing processes can be. Their relationship is ambivalent. There are numerous indications that the two of them are very fond of each other. For example, when Mama returned to the group after an absence of over a month, she spent hours grooming Nikkie and not Gorilla, Jimmie, Yeroen, or any of the other individuals with whom she normally spends her time. And of all the children in the colony, Moniek, Mama's daughter, is obviously Nikkie's favorite. But for a while it was the hostile side of their relationship that got the upper hand. This was at the beginning of Nikkie's leadership. Yeroen used to mobilize adult females against the young leader, and Mama was his major ally. At the end of such incidents, when Nikkie had been reconciled with Yeroen, he would go over to Mama to punish her for the part she had played. This could take a very long time, because Mama usually punished Nikkie in return by rejecting his subsequent attempts at reconciliation. For instance, Nikkie slaps Mama, but a little later he comes back and sits down by her, "shyly" plucking at some wisps of grass. Mama pretends she has not seen him, gets up and walks off. Nikkie waits a while, then starts all over again, with his hair on end. This was clearly a phase of negative reciprocity.

As Yeroen's resistance to Nikkie decreased, Mama became more favorably inclined toward Nikkie. She still supported Yeroen, but when Nikkie made his peace with her later she no longer took any "affective revenge," and their conflict remained brief. Later still—a process taking years—Mama reconciled her differences with Nikkie *before* his conflict with Yeroen had ended. One moment the two older apes were chasing after Nikkie; the next moment Mama affectionately embraced him. The conflict then continued between the two males, but Mama declined to take any further part.

In time, the situation became even stranger. Nikkie began kissing Mama before or even during his display against Yeroen. This developed gradually from their reconciliations, until it took place without any preceding conflict. It could be seen as a mark of Mama's neutrality. Nikkie and Mama were showing positive reciprocity.

I have done a statistical study of the bilateral nature of coalitions by comparing how each individual intervenes in the conflicts of the others. In periods of stability such interventions are symmetrical, both in a positive sense (two individuals support each other) and in a negative sense (two individuals support each other's opponents). If we are to get a full picture of reciprocity, however, we will have to analyze more kinds of behavior. Interventions need not necessarily be offset by other interventions. The receipt of regular support may be answered by greater tolerance toward the supporter, or by grooming. Perhaps we will eventually be able to conduct such an analysis in Arnhem. For the time being I should like to sum up as follows: chimpanzee group life is like a market in power, sex, affection, support, intolerance, and hostility. The two basic rules are "one good turn deserves another" and "an eye for an eye, a tooth for a tooth."[15]

The rules are not always obeyed, and flagrant disobedience may be punished. This happened once after Puist had supported Luit in chasing Nikkie. When Nikkie later displayed at Puist she turned to Luit and held out her hand to him in search of support. Luit, however, did nothing to protect her against Nikkie's attack. Immediately Puist turned on Luit, barking furiously, chased him across the enclosure, and even hit him. If her fury was in fact the result of Luit's failure to help her after she had helped him, this would suggest that reciprocity among chimpanzees is governed by the same sense of moral rightness and justice as it is among humans.

CONCLUSION

NOT ALL THE RESULTS OF THIS STUDY HOLD GOOD FOR CHIMPANZEES in general, because the rules governing social life partly depend on the living conditions and history of a group. Every community develops its own social traditions. On the other hand, such variations always revolve around certain fundamental themes that are characteristic for the species. The themes in our colony are no different from those in other chimpanzee groups. The reason why the Arnhem project has uncovered complexities unknown from studies in the natural habitat is simply that we could look at these apes in much greater detail.

FORMALIZATION.

Ranks are formalized. When they become unclear a dominance struggle ensues, after which the winner refuses reconciliation as long as his new status is not formally recognized.

INFLUENCE.

An individual's influence on group processes does not always correspond to his or her rank position. It also depends on personality, age, experience, and connections. I regard our oldest male and oldest female as the most influential group members.

COALITIONS.

Interventions in conflicts serve either to help friends and relatives or to build up powerful positions. The second, opportunistic type of intervention is seen specifically in the coalition formation of adult males and goes hand in hand with isolation tactics. There is evidence for a similar sex difference in humans.

BALANCE.

In spite of their rivalry, males form strong social bonds among themselves. They tend to develop a balanced power system based on their coalitions, individual fighting abilities, and support from females.

STABILITY.

Relationships among females are less hierarchically organized and much more stable than among males. A need for stability is also reflected in

Left to right, *Amber, Tarzan, Tepel, Gorilla.*

Thirsty chimpanzees catching rainwater dripping from a roof: left, Jimmie; right, Tepel.

the females' attitude toward male status competition. They even mediate between males.

EXCHANGES.

The human economic system, with its reciprocal transactions and centralization, is recognizable in the group life of chimpanzees. They exchange social favors rather than gifts or goods, and their support flows to a central individual who uses the prestige derived from it to provide social security. This is his responsibility, in the sense that he may undermine his own position if he fails to redistribute the support received.

MANIPULATION.

Chimpanzees are intelligent manipulators. Their ability is clear enough in their use of tools, but it is even more pronounced in their use of others as social instruments.

RATIONAL STRATEGIES.

It is possible that chimpanzees plan their dominance strategies before-hand. Although there is no proof of this, experimental studies suggest that we should keep an open mind on this question.

PRIVILEGES.

As a rule high-ranking males copulate more often than subordinates. This makes the evolution of male ambitions understandable, provided that mating success translates into reproductive success.

To my eyes, the most striking result is that there seem to be two layers of social organization. The first layer we see is a clear-cut rank order, at least among the most dominant individuals. Although primatologists spend a lot of energy discussing the value of the "dominance concept," they all know that it is impossible to ignore this hierarchical structure. The debate is not about its existence but about the degree to which knowledge of rank relationships helps to explain social processes. I think that, so long as we concentrate on the formal hierarchy, the explanations will be very poor indeed. We should also look behind it, at the second layer: a network of positions of influence. These positions are much more difficult to define, and I consider my descriptions in terms of influence and power only as imperfect first attempts. What I have seen, though, is that individuals losing a top rank certainly do not fall into oblivion: they are still able to pull many strings. In the same way, an individual rising in rank and at first sight appearing to be the big boss does not automatically have the greatest say in all matters. If it is hard to explain this duality of the social organization without using human terms, it is because we have very similar behind-the-scenes influences in our own society.

When Aristotle referred to man as a *political animal* he could not know just how near the mark he was. Our political activity seems to be a part of an evolutionary heritage we share with our close relatives. If someone had said this to me before I started working at Arnhem, I would have rejected such an idea as just too neat an analogy. What my work at Arnhem has taught me, however, is that the roots of politics are older than humanity. Nor would it be correct to accuse me of having, either consciously or unconsciously, projected human patterns onto chimpanzee behavior. The reverse is nearer the truth; my knowledge and experience of chimpanzee behavior have led me to look at humans in another light.

If we broadly define politics as social manipulation to secure and maintain influential positions, then politics involves every one of us. Outside central and local government we encounter this phenomenon in our family, at school, at work, and in meetings. Every day we cause conflicts or are party to those of others. We have both supporters and rivals, and we cultivate useful connections. These daily dabblings in politics are not always recognized as such, however, because people are past masters in camouflaging their true intentions. Politicians, for example, are vociferous about their ideals and promises but are careful not to disclose personal aspirations for power. This is not meant to be a reproach, because after all everyone plays the same game. I would go further and say that we are largely unaware that we are playing a game and hide our motives not only from others but also underestimate the immense effect they have on our own behavior. Chimpanzees, on the other hand, are quite blatant about their "baser" motives. Their interest in power is not greater than that of humanity; it is just more obvious.

Nearly five centuries ago Machiavelli described the political manipulations of the Italian princes, popes, and influential families such as the Medici and the Borgias without equivocation. Unfortunately his admirably realistic analysis has often been mistaken for a moral justification of these practices. One reason for this was that he presented rivalries and conflicts as constructive and not negative elements. Machiavelli was the first person to refuse to repudiate or cover up power motives. This violation of the existing collective lie was not kindly received. It was regarded as an insult to humanity.

To compare humans with chimpanzees can be taken to be just as insulting, or perhaps even more so, because human motives seem to become more animal as a result. And yet, among chimpanzees, power politics are not merely "bad" or "dirty." They give to the life of the Arnhem community its logical coherence and even a democratic structure. All parties search for social significance and continue to do so until a temporary balance is achieved. This balance determines the new hierarchical positions. Changing relationships reach a point where they become "frozen" in more or less fixed ranks. When we see how this formalization takes place during reconciliations, we understand that the hierarchy is a *cohesive* factor, which puts limits on competition and conflict. Child care, play, sex, and cooperation depend on the resultant stability. But underneath the surface the situation is constantly in a state of flux. The balance of power is tested daily, and if it proves too weak it is challenged and a new balance established. Consequently chimpanzee politics are also constructive. Humans should regard it as an honor to be classed as political animals.

Conclusion

EPILOGUE

My account ends in 1979, but ethological observations in Arnhem have continued, and the chimpanzees themselves have obviously never stopped politicking. A most dramatic event took place while I was still there, in 1980, but I decided against recounting it in the first edition of *Chimpanzee Politics* so as to avoid ending the book on a dark note. Moreover, at the time I was not yet emotionally ready to analyze this shocking event. I saved the story for my second book, in which it served as a reminder of how badly chimpanzees need reconciliation.

In the summer of 1980, a period of increased intolerance by Nikkie, the alpha male, led to a sudden breakup with Yeroen. Nikkie refused to let Yeroen copulate with estrus females. After several serious outbursts between the two, Yeroen withdrew his support of the leader. Overnight, Luit filled the power vacuum. With Nikkie groveling in the dust for him every day, Luit was a magnificent alpha male once again. This event demonstrated how much Nikkie had relied on Yeroen's support to remain at the top and also how closely the old male had monitored his end of the deal.

Luit was alpha for only ten weeks. The Yeroen–Nikkie alliance made a comeback with a bloody vengeance one night during which the two allies together severely injured Luit. Apart from biting off fingers and toes and causing deep gashes everywhere, the two aggressors removed Luit's testicles, which were found on the cage floor.[16] Luit died on the operating table due to loss of blood from the fight, which took place in a night cage with only the three senior males present. Given the victim's massive injuries and the relatively few injuries sustained by the other two, we must assume a remarkable level of coordination between Nikkie and Yeroen. Here is what happened the following day:

> We released Nikkie and Yeroen into the group. Immediately there
> occurred an unusually fierce attack on Nikkie by Puist. She was so
> persistently aggressive that Nikkie fled into a tree. On her own, Puist
> kept Nikkie there for at least ten minutes by screaming and charging
> each time he tried to come down. Puist had always been Luit's main
> ally among the females. She must have followed the fight for her

The Arnhem Zoo's breeding record is unsurpassed, and its chimpanzee colony has thrived despite deaths and removals. Tepel offers a rare glimpse of a newborn infant.

Today, the Arnhem colony is as lively and varied as ever, and it is not hard to recognize some old characters in the new faces. For example, we see Luit in Fons (above left) and Krom in Roosje (above right). Sabra (opposite), a daughter of Spin, resembles her natural sibling, Wouter. Sabra was raised by Jimmie after Spin's death in 1988. She now has a child of her own, a member of the rapidly growing third generation.

night cage offers a view into the males' pens. Later in the day, the group showed high interest in the two males, grooming and inspecting them. From that day on, Dandy played a much more important role than ever before. He repeatedly sought contact with Yeroen, resisting separation attempts by Nikkie. (from *Peacemaking among Primates*)

Thus, the day after the most gruesome attack in the history of the Arnhem colony, a new triangle emerged. Dandy, Nikkie, and Yeroen began repeating all of the dynamics seen within the original power trio in previous years. At around the same time, Toshisada Nishida published a paper describing a triangle among the chimpanzees in the Mahale Mountains in which an older male cunningly played off two younger, stronger males against each other. By regularly changing sides in their disputes, especially during periods of sexual rivalry, the older male made

the two others dependent upon him and increased his own mating success. The description very much reminded me of Yeroen's behavior. Nishida spoke of "allegiance fickleness" and suggested this might be a common strategy for post-prime males.

In 1983, Bert Haanstra, a nationally famous cineast, came to Arnhem to film the chimpanzees. Having read my book, he expected lots of political machinations, but unfortunately this was a time of relative stability, with Nikkie firmly in control. Undeterred, Haanstra decided to film for an entire summer, day in and day out. His patience paid off. The wonderful product, *The Family of Chimps*, presented the apes as real personalities, capturing chimpanzee social intelligence as no documentary had ever done before. It was a television hit all over the world. I had left the Netherlands before the movie was made and watched it the first time with tears in my eyes because of the loving attention with which all of my old friends had been depicted.

The following year, in 1984, Yeroen and Dandy were moving ever closer to a coalition against Nikkie. Yeroen had ceased to support Nikkie and increasingly resisted Nikkie's attempts to keep him away from Dandy. Nikkie must have been on edge like never before, because one morning, hearing screaming and hooting chimpanzees behind him, he ran out of the building at full speed, straight toward the moat surrounding the island. Almost exactly a year before, Nikkie had managed to cross the moat thanks to a thin layer of ice. Perhaps he thought that he could repeat this feat. This time, however, there was no ice, and Nikkie drowned then and there. The newspapers dubbed it a "suicide," but more likely it was a panic attack with a fatal outcome.

With Nikkie's death, the closeness between Yeroen and Dandy evaporated. Rivalry took its predictable place. Dandy became the new alpha, but the ghost of Nikkie still lingered, as revealed by the colony's striking reaction when *The Family of Chimps* was shown. One evening, in 1985, the winter hall was turned into a theater. With all the lights dimmed, the movie was projected on a light-colored wall. The apes watched in complete silence, some with their hair fully erect. When, in the movie, a female chimpanzee was attacked by pubertal males, several indignant barks were heard, but it remained unclear if the apes actually recognized the actors. Until, that is, Nikkie appeared. Dandy bared his teeth in a big, nervous grin and ran screaming to Yeroen: he embraced Yeroen and literally sat in his lap. Yeroen, too, had an uncertain grin on his face.

There was no doubt that both males had recognized the late leader. As explained by my successor, Otto Adang, Nikkie's "resurrection" had temporarily restored their old alliance!

In the years that followed, several key figures died of natural causes. This included Yeroen, Krom, and Spin. Others moved to faraway zoos, including several young males born in the early years of the colony as well as Henny and Puist, with their offspring. Given that the Arnhem colony has been the world's most successful chimpanzee-breeding group —with a total of seventy-five births and counting—death and removal never posed a threat to the colony, which still includes over thirty individuals.

Apart from the four adult males in my account, alpha males in Arnhem have included Tarzan, Fons, and Jing, a younger brother of Jonas. In general, however, the females have shown themselves better survivors than the males. With Dandy's recent death of heart failure in 1994, all original males are gone. But when the zoo celebrated the colony's twenty-five-year existence in 1996, Mama, Gorilla, Amber, Jimmie, Tepel, and Zwart were still around for the festivities. And there is also a new generation coming up, such as the offspring of Roosje and Moniek.

I regularly return to the Netherlands to see family and friends, including my old ape buddies. I visit them about once per year, and I am still recognized by the older generation. Mama always moves her old arthritic bones to the moat to greet me with pant-grunts, and Gorilla is perhaps the happiest of all. Ever since I taught her to bottle-feed Roosje, we have enjoyed a special bond. And each time I see the chimpanzees, I still feel grateful for having been at the right place at the right time to see a drama unfold which allowed me to question received dogma about the origins of politics.

NOTES

1. Initial reviews by political scientists can be found in *Politics and the Life Sciences* 2: 204–13 (1984), and Glendon Schubert (1986).

2. Journalists have exploited the account of the power struggles at the Arnhem Zoo for political ends by comparing national politicians with Nikkie, Luit, and Yeroen. This tendency has been particularly pronounced in the French media after Éditions du Rocher decided, in 1987, to place François Mitterrand and Jacques Chirac on the cover of *Chimpanzee Politics* with a grinning chimpanzee between them. The irreverent cover served to denigrate the politicians rather than to elevate the apes. A similar dilution of the book's message was reflected in the title under which Harnack Verlag, in 1983, published its German edition: *Unsere haarigen Vettern* ("Our Hairy Cousins"). These marketing decisions missed the point of my book, which is not to make fun of political leaders or apes, but to claim fundamental similarities and thus make people reflect on their own behavior.

3. Heini Hediger, a Swiss zoologist and ethologist, is widely regarded as the founder of *zoo biology,* a discipline seeking to understand the basic needs of captive animals so as to create settings in which a species' typical behavior can find expression. Modern zoos have moved away from keeping as many different species as possible to keeping fewer species in more spacious enclosures. The chimpanzee exhibit of the Arnhem Zoo represents one milestone on the long road from apes in small cages or dressed up for "tea parties" to apes in naturalistic enclosures. This development comes at the same time that apes in the wild are becoming endangered to the point that some populations now effectively live in sanctuaries under human protection and with occasional veterinary care. Thus, life in the natural habitat is beginning to resemble that at enlightened zoos.

4. Similarities between the Arnhem chimpanzees and chimpanzees in the wild are particularly striking in the studies of Toshisada Nishida in the Mahale Mountains of Tanzania. Many of the central concepts of the present analysis, such as separating interventions, side-directed behavior, and power-oriented coalition strategies, have proven useful for the understanding of hierarchical upheavals among the Mahale chimpanzees (e.g. Nishida and Hosaka, 1996). Other striking parallels have been drawn by Christopher Boehm (1994), who compared pacifying interventions among chimpanzees in Gombe National Park and at the Arnhem Zoo.

Both of these comparisons focused on male behavior. With regard to females there exists greater variation across chimpanzee communities. The Arnhem females seem to differ from their wild counterparts in that they are both more sociable and more politically influential (de Waal, 1994).

5. The Social Intelligence Hypothesis was developed in the 1950s and 1960s by Hans Kummer and Alison Jolly. At the Zurich Zoo, in Switzerland, Kummer observed how female hamadryas baboons enlist the support of an adult male against a female rival. The aggressor maneuvers herself between the male and her opponent, presenting her hindquarters to him while loudly threatening her. This behavior became known as *gesicherte Drohung,* or protected threat (Kummer, 1957). This was the first indication that primates do more than merely join each other in aggressive encounters; they seem to actively recruit support from others. Kummer (1971, p. 36) explained the enormous complexity of all this:

> In turning from one to another context at a rapid rate, the individual primate constantly adapts to the equally versatile activities of the group members around him. Such a society requires two qualities in its members: a highly developed capacity for releasing or suppressing their own motivations according to what the situation permits and forbids; and an ability to evaluate complex social situations, that is, to respond not to specific social stimuli but to a social field.

The idea that the evolution of primate intelligence took place under pressure of the social as opposed to the physical environment was developed most explicitly by Jolly (1966). Nick Humphrey (1976) also proposed a connection between social complexity and primate intelligence and contrasted social and technical problem-solving even more than Jolly had done.

When I began my studies at Arnhem, in 1975, I was extensively familiar with Kummer's ideas and already a great admirer of his work. Like him, I was intrigued by the tactical moves of primates to solicit and gain support from others. Previously, I had studied such behavior in long-tailed macaques under the supervision of Jan van Hooff at the University of Utrecht. The Arnhem chimpanzees showed a much greater variety of tactics, however, which so impressed and puzzled me that I was soon reading Niccolò Machiavelli for inspiration. But even though I take responsibility for introducing primatology to the Florentine chronicler of human nature, I have never felt comfortable with the label "Machiavellian intelligence" proposed, in 1988, by Richard Byrne and Andrew Whiten for social cognition in general.

Rightly or wrongly, the term *Machiavellian* implies a cynical, the-ends-justify-the-means exploitation of others. Social cognition covers much more than this. A mother resolving a weaning conflict by cleverly distracting her offspring, or an adult male waiting for the right moment to reconcile with his rival, both intelligently use their experience but are not exactly acting "Machiavellian" in the usual sense. Sensitivity to others, conflict resolution, and reciprocal exchange all demand a great deal of intelligence but are left out if our terminology one-sidedly emphasizes one-upmanship.

6. The effectiveness of high-ranking males in controlling aggression became clear a couple of years later when we tried to keep males and females

separate in the two large winter halls. We thought this might reduce tensions in the colony, as the males would have no females to compete over and the females and their offspring would be relieved of the frequent displays among males. After several weeks, the males were doing fine in their hall, but we noticed more and more fights among the females. One day it got so bad, with serious biting, that we had to send in the males when the females did not respond to our yelling. The males, who had followed the confrontation by ear, rushed into the female hall and broke up the fight right away. We had to repeat this maneuver several days later, with the same effect. I had never seen female chimpanzees go at each other like that. To prevent further injuries, we decided to keep the entire colony together.

7. The first observed mating with Puist occurred on 28 January 1981. This dramatic change of events was brought about by Nikkie. In the months before the first mating, he sexually invited Puist many times, and if she refused, he would start an elaborate bluff display that usually culminated in Puist becoming extremely agitated and going after Nikkie. Yeroen was a supporter of Puist during this period, but it was the large female herself who would carry out actual attacks on Nikkie. She injured him several times, showing that her resistance was serious and fierce. Nikkie was persistent, however, which resulted at first in brief presentations by Puist. She would present her behind, Nikkie would mount her, but she would jump away before intromission could occur. With time and continued pressure, the duration of Puist's presentations increased, until copulations took place. About one year later, Puist gave birth to a healthy daughter, Ponga. She turned out to be a perfect mother.

8. The most detailed descriptions and discussions of intercommunity warfare among chimpanzees can be found in Jane Goodall (1986) and Richard Wrangham and Dale Peterson (1996). Chimpanzee territoriality may even need to be taken into account by conservationists, as gathered from a recent newspaper account of thousands of chimpanzee deaths in Gabon. Apparently, the noise and traffic associated with mechanized logging pushed chimpanzees out of their forest ranges over a continuous front three to six miles wide. Biologist Lee White is quoted by William Stevens (1997) as speculating, on the basis of indirect evidence, that this may have triggered massive aggression when apes fled into the territory of the next community: "When that happens, you're essentially going to kick-start a chimpanzee war. The males from the invaded community attack the interlopers, and many die. Then the loggers keep coming. The invaded community itself is displaced onto the next community's territory. New warfare breaks out, and this process goes on and on and on and on as the loggers move through."

For details on bonobo social life see Takayoshi Kano (1992) and my book *Bonobo: The Forgotten Ape* (1997a). Bonobos are much less belligerent, both in captivity and in the field, than chimpanzees: they even engage in peaceful (and sexual) intergroup mingling.

9. An unwritten law allowing the lord to bed the newly wed wife of his serf and spend the first night with her. Sometimes only symbolic use was made of this right; the lord would put a leg into the bride's bed or he would climb on to the bed and pass over her.

10. The very first indication of infanticide in chimpanzees was an observation by Akira Suzuki (1971) of a large adult male in Budongo Forest holding a partly eaten dead infant of his species. The carcass was passed around among several males. Many more indications of infanticide in wild chimpanzees followed (reviewed by Russell Tuttle, 1986, pp. 122–24), including an actually filmed incident in the Mahale Mountains. The behavior is also known in a wide range of other species, from lions to prairie dogs, and from mice to langur monkeys. It is assumed that infanticidal males reduce their waiting time to fertilize a female: they eliminate the offspring of rivals so that the female starts cycling again. If the genes of infanticidal males spread faster than those of non-infanticidal males, the trait will be favored by natural selection (Sarah Blaffer Hrdy, 1979). According to this theory, males should target infants that they did not father themselves: victims are indeed usually offspring of strange females.

Had Nikkie wanted to kill Roosje? Since Roosje was in human care for a while, it may have looked as if she came from outside the colony. Whereas Krom seemed to realize that this was not the case, Nikkie may not have made the connection between the removed and the returned baby. If so, his response may have been the natural response of a male chimpanzee to a newborn unlikely to be his own, and we were lucky that Yeroen and Luit were there to stop him.

11. Asking approval before carrying out a particular action is of interest in relation to a possible moral order among chimpanzees (discussed in my book, *Good Natured,* 1996). Toward the end of Bert Haanstra's film, *The Family of Chimps,* there is a scene of young Wouter at the point of climbing a tree but first holding out his hand toward Dandy as if begging the adult male to let him go. Jane Goodall (1968, p. 281) describes a similar usage of the gesture by Melissa: "Sometimes in the early days of the feeding area we hid bananas in the trees so that the young chimpanzees could find them while the others fed from the boxes. One female invariably held her hand out towards the highest ranking male in the group several times before moving off to get a hidden banana that she had noticed."

12. A concept related to triadic awareness is Dorothy Cheney and Robert Seyfarth's (1990, pp. 72–86) "non-egocentric social knowledge," which refers to the fact that monkeys and apes learn aspects of social relationships in which they themselves are not directly involved, such as the hierarchy among others, or the matrilines to which other group members belong. This term emphasizes the ability of A to observe interactions between B and C and evaluate the B–C relationship, whereas the term triadic awareness considers how A needs to understand the B–C relationship because of its implications for

the A—B and A—C relationships. Experimental evidence for triadic awareness was provided by Verena Dasser (1988), who gave monkeys the task of classifying slides of other monkeys on the basis of what they knew about the social relationships between the depicted individuals.

13. Female-female relations are probably the most variable element of chimpanzee social organization. These relations range from close in all captive colonies known to me to rather loose in the wild populations at Gombe and Mahale, in Tanzania (e.g., Jane Goodall, 1986). But there exists variability in the wild as well. Female bonding appears to exist in a small chimpanzee population "trapped" by agricultural encroachment in a forest of approximately 6 km² on top of a mountain in Bossou, Guinea. Yukimaru Sugiyama (1984) often saw the majority of individuals in this forest travel together in a single party, and he measured relatively high rates of female-female grooming. Similarly, female chimpanzees seem more sociable in Taï Forest, in Ivory Coast, than at other sites: they frequently associate, develop special friendships, share food, and support one another. Christophe Boesch (1991) attributes this to cooperative defense against leopards. It is useful here to think in terms of *adaptive potentials* (de Waal, 1994): female chimpanzees have a distinct potential for bonding among themselves, but in most of their natural habitat this potential may not be realized due to the ecological pressure to disperse.

14. Group formation in captivity and studies in the wild have since thrown new light on female chimpanzee ambitions. At the Detroit Zoo serious tensions arose between newly introduced females, such that Kate Baker and Barbara Smuts (1994, p. 240) concluded: "When females are first forming relationships with one another, they use a variety of complex competitive strategies reminiscent of the strategies documented for status-striving males. These results . . . challenge previous characterizations of females as inherently less competitive than males."

Detailed records over the past three decades on the Gombe chimpanzees demonstrate that dominance, as measured by the direction of pant-grunts among females, has a dramatic effect on reproduction. The offspring of high-ranking females have better survival chances and mature faster than the offspring of low-ranking females. This means that dominance matters a great deal for wild female chimpanzees: probably, high rank translates into home ranges with the best quality foods (Pusey et al., 1997).

In captivity, food is plentiful for all members of the colony. My own experience with introductions remains that female chimpanzees quickly settle their dominance, usually without fighting. But even if the situation at the Detroit Zoo was exceptional, with two females claiming the top spot in the colony, the ensuing jockeying for position demonstrates an important female potential that one would never have suspected from watching a well-established, stable hierarchy such as existed in Arnhem during my study.

15. Since the appearance of *Chimpanzee Politics,* I have devoted a great deal

of research to reciprocal exchange in chimpanzees and other primates, including alliances and food sharing, and the exchange of different "currencies" such as sex-for-food in bonobos and food-for-grooming in chimpanzees. Perhaps the most convincing demonstration of reciprocity concerned the chimpanzee colony at the Yerkes Primate Center, where we documented grooming in the morning hours before conducting a food trial. Two large bundles of branches and leaves would be thrown into the enclosure, and sharing could begin as soon as the bundles had been claimed by some of the adults (and we would make sure it wasn't always the same ones). Our data demonstrated that chimpanzee A had a better chance of getting food from B after A had groomed B. The grooming by A had no effect on sharing by A himself nor on sharing by B with others. The specificity of the exchange, which has thus far not been demonstrated for any other animal, means that chimpanzees keep track of received services, and that they return favors. See de Waal (1989b, 1997b), and de Waal and Lesleigh Luttrell (1988).

16. The incident with Luit is described in detail by de Waal (1986). Jane Goodall (1992) reported a mass attack on a male, Goblin, in Gombe National Park that also led to serious scrotal injuries. These injuries would most likely have cost Goblin's life had he not been treated by a veterinarian. A further similarity with the incident in Arnhem is that the attack on Goblin occurred within the same community, whereas lethal violence among wild chimpanzees typically involves males of different communities.

BIBLIOGRAPHY

Alexander, R. (1975). "The Search for a General Theory of Behavior." *Behavl. Sci.* 20: 77–100.

Asquith, P. (1984). "The Inevitability and Utility of Anthropomorphism in Description of Primate Behaviour." In *The Meaning of Primate Signals,* ed. R. Harré and V. Reynolds. Cambridge: Cambridge Univ. Press.

Baker, K. C., and B. B. Smuts (1994). "Social Relationships of Female Chimpanzees: Diversity between Captive Social Groups." In *Chimpanzee Cultures,* ed. R. W. Wrangham, W. C. McGrew, F. B. M. de Waal, and P. Heltne. Cambridge: Harvard Univ. Press. Pp. 227–42.

van den Berghe, P. (1980). "Incest and Exogamy: A Sociobiological Reconsideration." *Ethol. Sociobiol.* 1: 151–62.

Bernstein, I. (1969). "Spontaneous Reorganization of a Pigtail Monkey Group." Proceedings 2nd Congress IPS, Atlanta 1968, vol. 1: 48–51. Basel: Karger.

Bernstein, I. (1976). "Dominance, Aggression and Reproduction in Primate Societies." *J. Theor. Biol.* 60: 459–72.

Bernstein, I., and L. Sharpe (1966). "Social Roles in a Rhesus Monkey Group." *Behaviour* 26: 91–103.

Bindra, D. (1976). *A Theory of Intelligent Behavior.* New York: Wiley. Pp. 313–19.

Boehm, C. (1994). "Pacifying Interventions at Arnhem Zoo and Gombe." In *Chimpanzee Cultures,* ed. R. W. Wrangham, W. C. McGrew, F. B. M. de Waal, and P. Heltne. Cambridge: Harvard Univ. Press. Pp. 211–26.

Boesch, C. (1991). "The Effects of Leopard Predation on Grouping Patterns in Forest Chimpanzees." *Behaviour* 117: 220–42.

Bond, J., and W. Vinacke (1961). "Coalitions in Mixed-sex Triads." *Sociometry* 24: 61–75.

Buss, D. M., R. J. Larson, D. Westen, and J. Semmelroth (1992). "Sex Differences in Jealousy: Evolution, Physiology, and Psychology." *Psych. Sci.* 3: 251–55.

Bygott, D. (1974). "Agonistic Behaviour in Wild Chimpanzees." Ph.D. thesis, Cambridge U.K. (unpublished).

Byrne, R., and A. Whiten, eds. (1988). *Machiavellian Intelligence.* Oxford: Clarendon.

Cheney, D. L., and R. M. Seyfarth (1990). *How Monkeys See the World: Inside the Mind of Another Species.* Chicago: Univ. of Chicago Press.

Dasser, V. (1988). "A Social Concept in Java Monkeys." *Anim. Behav.* 36: 225–30.

Dearden, J. (1974). "Sex-linked Differences of Political Behavior: An Investigation of Their Possibly Innate Origins." *Soc. Sci. Inform.* 13: 19–25.

Dennett, D. (1983). "Intentional Systems in Cognitive Ethology: The 'Panglossian Paradigm' Defended." *Behav. Brain Sci.* 6: 343–90.

Döhl, J. (1968). "Uber die Fähigkeit einer Schimpansin, Umwege mit selbst-ändigen Zwischenzielen zu überblicken." *Z. Tierpsychol.* 25: 89–103.

Döhl, J. (1970). "Zielorientiertes Verhalten beim Schimpansen." *Naturwissen-schaft und Medizin* 34: 43–57.

Freud, S. (1921). *Group Psychology and the Analysis of the Ego.* London: Hogarth, 1967.

Gallup, G. (1970). "Chimpanzees: Self-recognition." *Science* 167: 86–87.

Gamson, W. (1961). "A Theory of Coalition Formation." *Amer. Soc. Rev.* 26: 373–82.

Gardner, R., and B. Gardner (1969). "Teaching Sign-language to a Chimpan-zee." *Science* 165: 664–72.

Gardner, R., and B. Gardner (1977). "Comparative Psychology and Language Acquisition." Paper given at the XVth International Ethological Confer-ence in Bielefeld, W. Germany (unpublished).

Ginsburg, H., and S. Miller (1981). "Altruism in Children: A Naturalistic Study of Reciprocation and an Examination of the Relationship between Social Dominance and Aid-giving Behavior." *Ethol. Sociobiol.* 2: 75–83.

Goodall, J. van Lawick- (1968). "The Behaviour of Free-living Chimpanzees in the Gombe Stream Reserve." *Anim. Behav. Monograph* 3.

Goodall, J. van Lawick- (1971). *In the Shadow of Man.* London: Collins; Boston: Houghton Mifflin.

Goodall, J. van Lawick- (1975). "The Chimpanzee." In *The Quest for Man,* ed. V. Goodall. London: Phaidon.

Goodall, J. (1979). "Life and Death at Gombe." *Nat. Geogr.* 155: 592–621.

Goodall, J. (1986). *The Chimpanzees of Gombe.* Cambridge, Mass.: Belknap.

Goodall, J. (1992). "Unusual Violence in the Overthrow of an Alpha Male Chimpanzee at Gombe." In *Topics in Primatology: Vol. 1, Human Origins,* ed. T. Nishida, W. C. McGrew, P. Marler, M. Pickford, and F. B. M. de Waal. Tokyo: Univ. of Tokyo Press. Pp. 131–42.

Griffin, D. (1976). *The Question of Animal Awareness.* New York: Rockefeller Univ. Press.

de Groot, A. (1965). *Thought and Choice in Chess.* The Hague: Mouton.

Hall, K., and I. DeVore (1965). "Baboon Social Behavior." In *Primate Behavior,* ed. I. DeVore. New York: Holt.

Halperin, S. (1979). "Temporary Association Patterns in Free Ranging Chim-panzees; An Assessment of Individual Grouping Preferences." In *The Great Apes,* ed. D. Hamburg and E. McCown. Benjamin/Cummings, California.

Hausfater, G. (1975). "Dominance and Reproduction in Baboons (*Papio cyno-cephalus*); A Quantitative Analysis." *Contributions to Primatology* 7. Basel: Karger.

Hobbes, T. (1991 [1651]). *Leviathan.* Cambridge: Cambridge Univ. Press.

van Hooff, J. (1973). "The Arnhem Zoo Chimpanzee Consortium: An Attempt to Create an Ecologically and Socially Acceptable Habitat." *Int. Zoo Year-book* 13: 195–205.

van Hooff, J. (1974). "A Structural Analysis of the Social Behaviour of a Semi-captive Group of Chimpanzees." In *Social Communication and Movement,* ed. M. von Cranach and I. Vine. London: Academic Press.

Hrdy, S. B. (1979). "Infanticide among Animals: A Review, Classification, and Examination of the Implications for the Reproductive Strategies of Females." *Ethol. Sociobiol.* 1: 13–40.

Humphrey, N. K. (1976). "The Social Function of Intellect." In *Growing Points in Ethology,* ed. P. Bateson and R. A. Hinde. Cambridge: Cambridge Univ. Press. Pp. 303–21.

Isaac, G. (1978). "The Food-sharing Behavior of Protohuman Hominids." *Scientific American* 238: 90–108.

Jolly, A. (1966). "Lemur Social Behavior and Primate Intelligence." *Science* 153: 501–6.

Kano, T. (1992). *The Last Ape.* Stanford: Stanford Univ. Press.

Kaufmann, J. (1965). "A Three Year Study of Mating Behavior in a Free-ranging Band of Rhesus Monkeys." *Ecology* 46: 500–512.

Kawai, M. (1958). "On the System of Social Ranks in a Natural Troop of Japanese Monkeys." *Primates* 1: 111–48. English translation in *Japanese Monkeys,* ed. K. Imanishi and S. Altmann. Atlanta: Emory Univ. Press, 1965.

Köhler, W. (1917). *Intelligenzprüfungen an Menschenaffen.* Berlin: Springer, 1973. Translated as *The Mentality of Apes.* New York: Vintage Books, 1959.

Kolata, G. (1976). "Primate Behavior: Sex and the Dominant Male." *Science* 191: 55–56.

Kortlandt, A. (1969). "Chimpansees." In *Het Leven der Dieren,* ed. B. Grzimek, Band XI, pp. 14–49. Utrecht: Het Spectrum. P. 46.

Kropotkin, P. (1899). *Memoires van een Revolutionair.* Baarn, Netherlands: Wereldvenster, 1978. P. 314. Translated from *Memoirs of a Revolutionist.* New York: Dover, 1971.

Kropotkin, P. (1902). *Mutual Aid: A Factor of Evolution.* New York: New York University Press, 1972.

Kummer, H. (1957). *Soziales Verhalten einer Mantelpavian Gruppe.* Bern: Verlag Hans Huber.

Kummer, H. (1971). *Primate Societies.* Chicago: Aldine.

Lasswell, H. (1936). *Who Gets What, When, and How.* New York: McGraw-Hill.

Leakey, R., and R. Lewin (1977). *Origins.* London: Macdonald & Jane's; New York: Dutton.

Linton, R. (1936). *The Study of Man: An Introduction.* New York: Appleton, 1964. Student's edition, p. 184.

Lorenz, K. (1931). "Beiträge zur Ethologie Sozialer Corviden." In *Gesammelte Abhandlungen,* Band I: 13–69. Munich: Piper, 1965.

Lorenz, K. (1959). "Gestaltwahrnehmung als Quelle Wissenschaftlicher Erkenntnis." In *Gesammelte Abhandlungen,* Band II: 255–300. Munich: Piper, 1967.

Machiavelli, N. (1532). *The Prince.* In *The Portable Machiavelli,* ed. P. Bondanella and M. Musa. Harmondsworth: Penguin Books, 1979.

Maslow, A. (1936-7). "The Role of Dominance in Social and Sexual Behavior of Infra-human Primates." Series of articles in *J. Genet. Psychol.* 48 and 49.

Mauss, M. (1924). *The Gift: Forms and Functions of Exchange in Archaic Societies.* London: Routledge & Kegan Paul, 1974.

Menzel, E. (1971). "Communication about the Environment in a Group of Young Chimpanzees." *Folia primatol.* 15: 220-32.

Menzel, E. (1972). "Spontaneous Invention of Ladders in a Group of Young Chimpanzees." *Folia primatol.* 17: 87-106.

Mori, A. (1977). "The Social Organization of the Provisioned Japanese Monkey Troops which have Extraordinarily Large Population Sizes." *J. Anthrop. Soc. Nippon* 85: 325-45.

Morris, D. (1979). *Animal Days.* London: Jonathan Cape. P. 147.

Mulder, M. (1972). *Het Spel om Macht; over Verkleining en Vergroting van Machtsongelijkeid.* Meppel, Netherlands: Boom.

Mulder, M. (1979). *Omgaan met Macht.* Amsterdam: Elsevier.

Nacci, P., and J. Tedeschi (1976). "Liking and Power as Factors Affecting Coalition Choices in the Triad." *Soc. Behav. Personality* 4(1): 27-32.

Nadler, R. (1976). "Rann vs. Calabar: A Study in Gorilla Behavior." *Yerkes Newsletter* 13(2): 11-14.

Nadler, R., and B. Tilford (1977). "Agonistic Interactions of Captive Female Orang-utans with Infants." *Folia primatol.* 28: 298-305.

Nieuwenhuijsen, K., and F. de Waal (1982). "Effects of Spatial Crowding on Social Behavior in a Chimpanzee Colony." *Zoo Biol.* 1: 5-28.

Nishida, T. (1979). "The Social Structure of Chimpanzees of the Mahale Mountains." In *The Great Apes,* ed. D. Hamburg and E. McCown. Benjamin/Cummings, California.

Nishida, T. (1983). "Alpha Status and Agonistic Alliance in Wild Chimpanzees." *Primates* 24: 318-36.

Nishida, T., and K. Hosaka (1996). "Coalition Strategies among Adult Male Chimpanzees of the Mahale Mountains, Tanzania." In *Great Ape Societies,* ed. W. C. McGrew, L. F. Marchant, and T. Nishida. Cambridge: Cambridge Univ. Press. Pp. 114-34.

Noë, R., F. de Waal, and J. van Hooff (1980). "Types of Dominance in a Chimpanzee Colony." *Folia primatol.* 34: 90-110.

Pusey, A. (1980). "Inbreeding Avoidance in Chimpanzees." *Anim. Behav.* 28: 543-52.

Pusey, A., J. Williams, and J. Goodall (1997). "The Influence of Dominance Rank on the Reproductive Success of Female Chimpanzees." *Science* 277: 828-31.

Riss, D., and C. Busse (1977). "Fifty-day Observation of a Free-ranging Adult Male Chimpanzee." *Folia primatol.* 28: 283-97.

Riss, D., and J. Goodall (1977). "The Recent Rise to the Alpha Rank in a Population of Free-living Chimpanzees." *Folia primatol.* 27: 134-51.

Sahlins, M. (1965). "On the Sociology of Primitive Exchange." In *The Relevance* of *Models for Social Anthropology,* ed. M. Banton. A.S.A. Monograph 1. London: Tavistock.

Sahlins, M. (1972). "The Social Life of Monkeys, Apes and Primitive Man." In *Primates on Primates,* ed. D. Quiatt. Minneapolis: Burgess.

Sahlins, M. (1977). *The Use and Abuse of Biology.* London: Tavistock.

van de Sande, J. (1973). "Speltheoretische Onderzoekingen naar Gedrags-Verschillen Tussen Mannen en Vrouwen." *Nederl. T. Psychol.* 28: 327–41.

Schjelderup-Ebbe, T. (1922). "Beiträge zur Sozialpsychologie des Haushuhns." *Z. Psychol.* 88: 225–52.

Schubert, G. (1986). "Primate Politics." *Soc. Sci. Information* 25: 647–80.

Silk, J. (1979). "Feeding, Foraging, and Food-sharing Behavior of Immature Chimpanzees." *Folia primatol.* 31: 123–42.

Stevens, W. K. (1997, May 13). "Gabon Logging Pushes Chimps into Deadly Territorial War." *The New York Times.*

Sugiyama, Y. (1984). "Population Dynamics in Wild Chimpanzees at Bossou, Guinea, between 1976 and 1983." *Primates* 25: 391–400.

Sugiyama, Y., and J. Koman (1979). "Social Structure and Dynamics of Wild Chimpanzees at Bossou, Guinea." *Primates* 20: 323–39.

Suzuki, A. (1971). "Carnivority and Cannibalism Observed among Forest-Living Chimpanzees." *J. Anthrop. Soc. Nippon* 79: 30–48.

Teleki, G. (1973). *The Predatory Behavior of Wild Chimpanzees.* Lewisburg, Pa.: Bucknell Univ. Press.

Thibaut, J., and H. Kelley (1959). *The Social Psychology of Groups.* New York: Wiley. P. 37.

Trivers, R. (1971). "The Evolution of Reciprocal Altruism." *Q. Rev. Biol.* 46: 35–57.

Trivers, R. (1974). "Parent-offspring Conflict." *Am. Zool.* 14: 249–64.

Tutin, C. (1975). "Exceptions to Promiscuity in a Feral Chimpanzee Community." In *Contemporary Primatology,* 5th Congress IPS, Nagoya 1974, pp. 445–49. Basel: Karger.

Tutin, C. (1979). "Responses of Chimpanzees to Copulation: With Special Reference to Interference by Immature Individuals." *Anim. Behav.* 27: 845–54.

Tuttle, R. H. (1986). *Apes of the World: Their Social Behavior, Communication, Mentality, and Ecology.* Park Ridge, NJ: Noyes.

de Waal, F. B. M. (1975). "The Wounded Leader: A Spontaneous Temporary Change in the Structure of Agonistic Relations among Captive Java-monkeys (*Macaca fascicularis*)." *Netherlands' J. Zoology* 25: 529–49.

de Waal, F. (1977). "The Organization of Agonistic Relations within Two Captive Groups of Java-monkeys (*Macaca fascicularis*)." *Z. Tierpsychol.* 44: 225–82.

de Waal, F. (1978). "Exploitative and Familiarity-dependent Support Strategies in a Colony of Semi-free-living Chimpanzees." *Behaviour* 66: 268–312.

de Waal, F. (1980). "Schimpansin zieht Stiefkind mit der Flasche auf." *Das Tier* 20: 28–31.

de Waal, F. B. M. (1986). "The Brutal Elimination of a Rival among Captive Male Chimpanzees." *Ethol. Sociobiol.* 7: 237–51.

de Waal, F. B. M. (1989a). *Peacemaking among Primates.* Cambridge: Harvard Univ. Press.

de Waal, F. B. M. (1989b). "Food Sharing and Reciprocal Obligations among Chimpanzees." *J. Human Evol.* 18: 433–59.

de Waal, F. B. M. (1994). "The Chimpanzee's Adaptive Potential: A Comparison of Social Life under Captive and Wild Conditions." In *Chimpanzee Cultures,* ed. R. W. Wrangham, W. C. McGrew, F. B. M. de Waal, and P. Heltne. Cambridge: Harvard Univ. Press. Pp. 243–60.

de Waal, F. B. M. (1996). *Good Natured: The Origins of Right and Wrong in Humans and Other Animals.* Cambridge: Harvard Univ. Press.

de Waal, F. B. M. (1997a). *Bonobo: The Forgotten Ape* (with photographs by F. Lanting). Berkeley: Univ. of California Press.

de Waal, F. B. M. (1997b). "The Chimpanzee's Service Economy: Food for Grooming." *Evol. Human Behav.* 18: 1–12.

de Waal, F., and J. Hoekstra (1980). "Contexts and Predictability of Aggression in Chimpanzees." *Anim. Behav.* 28: 929–37.

de Waal, F., and L. Luttrell (1988). "Mechanisms of Social Reciprocity in Three Primate Species: Symmetrical Relationship Characteristics or Cognition?" *Ethol. Sociobiol.* 9: 101–18.

de Waal, F., and A. van Roosmalen (1979). "Reconciliation and Consolation among Chimpanzees." *Behav. Ecol. Sociobiol.* 5: 55–66.

Watanabe, K. (1979). "Alliance Formation in a Free-ranging Troop of Japanese Macaques." *Primates* 20: 459–74.

Wight, M. (1946). *Power Politics.* New ed., H. Bull and C. Holbraad, eds. Harmondsworth: Penguin Books, 1979.

Wrangham, R. (1974). "Artificial Feeding of Chimpanzees and Baboons in Their Natural Habitat." *Anim. Behav.* 22: 83–93.

Wrangham, R. (1975). "Behavioural Ecology of Chimpanzees in Gombe National Park, Tanzania." Ph.D. thesis, Cambridge U.K. (unpublished).

Wrangham, R. W., and D. Peterson (1996). *Demonic Males: Apes and the Origins of Human Violence.* Boston: Houghton Mifflin.

van Wulfften Palthe, T. (1978). "De Beschrijving van een Machtswisseling, 1973–74, bij de Chimpansees van Burgers' Dierenpark." Doctoral report (unpublished).

van Wulfften Palthe, T., and J. van Hooff (1975). "A Case of Adoption of an Infant Chimpanzee by a Suckling Foster Chimpanzee." *Primates* 16: 231–34.

Zinnes, D. (1970). "Coalition Theories and the Balance of Power." In *The Study of Coalition Behavior,* ed. S. Groennings, E. Kelley, and M. Leiserson. New York: Holt.

ACKNOWLEDGMENTS

IN A SENSE THIS STUDY IS THE RESULT OF THE STRONG DUTCH ETHO-
logical tradition. By this I do not mean the speculative comparisons of
humans and animals, which are entirely my own, but the method of
patient observation and meticulous recording. The ethologist who has
had the greatest effect on me is Jan van Hooff. I worked with him in
Utrecht for four years before I came to Arnhem in 1975. But also afterward, while I was studying chimpanzees, I was a member of the staff
of his university department. Consequently there are very few findings
and theoretical problems in this book which Jan and I have not discussed at length.

The author in 1980, when he was writing Chimpanzee Politics *(photo by Catherine Marin).*

I was introduced to chimpanzee behavior by two students, Jan Brinkhuis and Rob
Slaager. Later I coordinated chimpanzee
research at Arnhem, working with a whole
succession of students, about four per year.
The project inspired their enthusiasm
and produced accurate observations from
them. The constant discussions of events
in the colony have always been a great
stimulus to me. My thanks go to: Otto
Adang, Dirk Fokkema, Agaath Fortuyn
Droogleever, Aaltjen Grotenhuis, Ruud
Harmsen, Rob Hendriks, Janneke Hoekstra, Kees Nieuwenhuijsen, Ronald Noë,
Trix Piepers, Marieke Polder, Albert Rama-
kers, Angeline van Roosmalen, Claudia
Roskam, Fred Ruoff, and Mariëtte van der
Weel. And also to several students who preceded me at Arnhem: Joost
Meulenbroek, Ted Polderman, and Titia van Wulfften Palthe.

Our research was conducted under the auspices of the Laboratory
of Comparative Physiology of the University of Utrecht. The Laboratory gave us our literature, analyzed the students' findings, repaired our
equipment, and assisted us in many other ways. My grateful thanks,

therefore, go to all its staff and to the University, which helped to finance the project. The students and animal keepers at Arnhem, especially Jacky Hommes (who for the last seventeen years has taken care of the chimpanzees), are thanked for pointing out new and changed individuals each time I visit. For this revised edition, I am grateful to Frank Kiernan, photographer at the Yerkes Primate Center, for his expert assistance with the printing of my twenty-year-old negatives.

Unlike my current writing, which I do directly in English on a word processor, the original version of *Chimpanzee Politics* is a pencil-on-paper manuscript in my native Dutch. This manuscript was expertly translated by Janet Milnes. I am grateful to Desmond Morris and Tom Maschler for believing in me, stimulating me to write in a popular style, and helping me to reach an international audience by getting the book published in English. Finally, I would like to thank my wife, Catherine Marin. She helped me keep the book simple and direct. She also communicated to me her knowledge of photography, not to mention the love and support she gave me then, and gives me now.

230

Acknowledg-ments

INDEX

Library of Congress Cataloging-in-Publication Data

Waal, F. B. M. de (Frans B. M.), 1948–
    [Chimpansee politiek.   English]
Chimpanzee politics : power and sex among apes / Frans de Waal ; with photographs
and drawings by the author. — Rev. ed.
        p.   cm.
    Includes bibliographical references and index.
    ISBN 0-8018-5839-9 (alk. paper)
    1. Chimpanzees—Behavior.   2. Social behavior in animals.   3. Sexual behavior
in animals.   I. Title.
QL737.P96W3213   1998
599.885'156—dc21
                                                                                97-44284
                                                                                    CIP